BIOCHEMICAL FACTORS CONCERNED IN THE FUNCTIONAL ACTIVITY OF THE NERVOUS SYSTEM – A COLLECTION OF ABSTRACTS

BIOCHEMICAL FACTORS CONCERNED IN THE FUNCTIONAL ACTIVITY OF THE NERVOUS SYSTEM

Edited by

D. RICHTER

FIRST INTERNATIONAL MEETING

OF THE

INTERNATIONAL SOCIETY FOR NEUROCHEMISTRY

STRASBOURG 1967

THE QUEEN'S AWARD TO INDUSTRY 1966

PERGAMON PRESS

OXFORD : LONDON : EDINBURGH : NEW YORK

TORONTO : SYDNEY : PARIS : BRAUNSCHWEIG

Pergamon Press Ltd., Headington Hill Hall, Oxford
4 & 5 Fitzroy Square, London W.1

Pergamon Press (Scotland) Ltd., 2 & 3 Teviot Place, Edinburgh 1

Pergamon Press Inc., Maxwell House, Fairview Park, Elmsford, New York 10523

Pergamon of Canada Ltd., 207 Queen's Quay West, Toronto 1

Pergamon Press (Aust.) Pty. Ltd., 19a Boundary Street, Rushcutters Bay, N.S.W. 2011, Australia

Pergamon Press S.A.R.L., 24 rue des Écoles, Paris 5e

Vieweg & Sohn GmbH, Burgplatz 1, Braunschweig

First edition 1969

Library of Congress Catalog Card No. 75–80290

Printed in Great Britain by A. Wheaton & Co., Exeter

08 013311 8

BIOCHEMICAL FACTORS CONCERNED IN THE FUNCTIONAL ACTIVITY OF THE NERVOUS SYSTEM

INTERFACIAL INTERACTION OF CALCIUM AND ATP WITH PHOSPHOLIPIDS AND OTHER SURFACTANTS

Leo G. Abood

Center for Brain Research, University of Rochester, Rochester, New York.

One of the initial events in nerve conduction is the rapid change in the permeability of the excitatory membrane to external Na^+. It has been proposed that the permeability change results from a dissociation of Ca^{2+} from a macromolecular complex comprised of phospholipids, polysaccharides, and proteins. To pursue this hypothesis studies have been carried out with surface monolayers of lipids and amphiphiles measuring such parameters as surface pressure, surface potential, and surface adsorption (radioactive tracer technique). Ca^{2+} interaction with monolayers of phospholipids and acidic amphiphiles readily occurs. Although stable Ca-phospholipid-ATP complexes are not readily demonstrable in the monolayer, in the presence of cationic soaps, lipophilic drugs, and other lipophilic substances it is possible to demonstrate an interfacial interaction involving Ca^{2+} and ATP. Electron micrographs of calcium-ATP-phospholipid complexes will be presented. More complex systems, involving surface micelles and other aggregates, with amphiphiles no longer conform to Gibbsian adsorption but probably involve ternary liquid crystal structures. The demonstration of a sequestering and chelating action of ATP on Ca-lipid complexes in surface films supports the notion that ATP plays a regulatory role in excitatory phenomena.

IONIC PROPERTIES OF ACIDIC LIPIDS: PHOSPHATIDYL INOSITOL

M.B. Abramson, G. Colacicco and M.M. Rapport

The Saul R. Korey Department of Neurology and the Department of Biochemistry, Albert Einstein College of Medicine, New York.

Ion binding and other physical properties of phosphatidyl inositol (PI) in aqueous media were studied in relation to the role that acidic lipids play in biological systems.

Clear dispersions of 8-10 mg NH_4-PI or Na-PI in 5 ml of water, obtained by gentle sonic radiation, were titrated in the pH range of 2.5 to 10.0. The initial pH was 6.1 to 6.5. The pK' of PI, which was 3.25 in the absence of electrolyte, was lowered to 2.50 in 0.08 M-NaCl.

The interaction of PI with metallic cations was studied by three methods: H^+ release, turbidity, and coagulation. (1) Dispersions of PI were adjusted to pH 3.5 and increments of electrolyte were added. The quantity of H^+ released (measured by the quantity of tetramethylammonium hydroxide required to maintain pH 3.5) followed the order Ca > Mg >> K > Na. (2) Dispersions of PI were buffered at pH 7.3 and the turbidity was measured by the scattering ratio $I_{90°}/I_{0°}$. The effectiveness of the cations in increasing turbidity was $Ca^{++} > Mg^{++} >> K^+ > Na^+ > Li^+ >$ choline chloride. (3) Dispersions of PI coagulated on the addition of salts. Analysis of cations in the coagula provided a measure of cation selectivity. The molar ratio of divalent cation to P of 0.5 indicated bridging of two PI molecules by each Ca^{++} or Mg^{++}. The selectivity ratios were $Ca^{++} \simeq 2$ and $K^+ \simeq 1.2$.

The viscosity of PI dispersions in water was much greater than for other acidic lipids; however, a low concentration (2.5 mM) of NaCl produced a sharp decrease. In PI binding the order of effectiveness for divalent cations ($Ca^{++} > Mg^{++}$) was similar to that found with sulphatide, phosphatidic acid (PA), and phosphatidyl serine (PS). The order with univalent cations ($K^+ > Na^+ > Li^+$) was similar with PI and sulphatide, but the reverse of that with PA and PS. The observed differences depend on polarizability, charge density, and hydration.

Supported by grants from the Office of Saline Water No. 14-01-0001-635 and the American Cancer Society No.P410.

EFFECT OF VARIOUS AGENTS ON PROTEIN AND RNA SYNTHESIS IN THE GOLDFISH BRAIN

Bernard W. Agranoff, Ramon Lim, Luigi Casola and John J. Brink

Mental Health Research Institute, University of Michigan, U.S.A.

Previous studies in this laboratory have demonstrated that the formation of long-term memory is blocked by puromycin injected intracranially immediately before or after a training session. Such injections blocked the incorporation of labelled [4,5-^{3}H]leucine into brain protein for several hours but had no detectable effect on [5-^{3}H]uridine incorporation into whole brain RNA. In the absence of the drug, goldfish showed an increased incorporation of labelled leucine into protein during training. Like puromycin, acetoxycycloheximide blocked memory formation and protein synthesis, but not RNA formation. Further studies with various antimetabolites support the concept of an obligatory protein synthetic step in long-term memory formation. Pemoline (2-imino-4-keto-5-phenyloxazolidine) and U-9189 (1,1,3-tricyano-2-amino-1-propene), two agents reported to stimulate RNA synthesis in mammalian brain, had no effect on labelled uridine incorporation into whole brain RNA in the goldfish. Following a single intracranial injection of 2 μg of actinomycin D, RNA formation was rapidly depressed, while a decrease in protein synthesis developed several hours following the injection.

(Supported by grants from the National Institutes of Health and National Science Foundation).

THE INFLUENCE OF NERVE IMPULSE FLOW ON THE EFFECTS OF DRUGS ON CENTRAL MONOAMINE NERVE TERMINALS

Nils-Erik Andén, Hans Corrodi, Kjell Fuxe and Tomas Hokfelt

Department of Pharmacology, University of Göteborg, Göteborg and Department of Histology, Karolinska Institutet, Stockholm.

Noradrenaline (NA) and 5-hydroxytryptamine (5-HT) of the spinal cord are present in axon terminals of neurons whose cell bodies occur in the lower brain stem. Thus, a spinal cord transection removes the influence of the action potentials on the monoamine nerve terminals lying caudal but not cranial to the lesion. Both biochemical and histochemical studies on sectioned rat spinal cords have demonstrated two types of monoamine-reducing drugs: those dependent on nervous impulse flow and those primarily independent. The inhibitors of monoamine biosynthesis belong to the former group. The NA of the cranial but not of the caudal part of the sectioned cord was, thus, considerably reduced after inhibition of the tyrosine hydroxylase (α-methyl-p-tyrosine, α-propyldopacetamide), or the dihydroxyphenylalanine decarboxylase (seryltrihydroxybenzylhydrazine), or the dopamine-β-oxidase (diethyldithiocarbamate). Similar observations were made concerning 5-HT after inhibition of the tryptophan hydroxylase (α-propyldopacetamide) or the 5-hydroxytryptophan decarboxylase (seryl-trihydroxybenzylhydrazine). In intact spinal cords all these drugs produced the same reduction of NA and 5-HT throughout the spinal cord. On the other hand, the reductions of NA and 5-HT after inhibition of the uptake-storage mechanism by reserpine or tetrabenazine were about equal cranial and caudal to a transection, suggesting that the amine movements from the granules to the cytoplasm are not influenced by the action potentials.

After inhibition of the monoamine oxidase the concentrations of NA and 5-HT in the spinal cord increased about three times more cranial than caudal to a transection, whereas in intact cords the increases were about equal in the two parts, indicating a stimulation of synthesis by the nerve impulses. Such a synthesis stimulation probably compensates for the losses occurring when the amines are released from the nerve terminals by the action potentials.

CARBOHYDRATE METABOLISM IN EMBRIONIC SENSORY GANGLIA. EFFECT OF A SPECIFIC PROTEIN "NERVE GROWTH FACTOR" (NGF).

P.U. Angeletti, A. Liuzzi and F. Pocchiari

Laboratori di Chimica Biologica,
Istituto Superiore di Sanitá, Roma, Italy.

It was previously shown in this Laboratory that the addition of NGF greatly stimulates the oxidative process in sensory ganglia incubated _in vitro_ (ANGELETTI _et al_., 1964).

In the present report quantitative date are presented on the effect of NGF on the general pattern of glucose and pyruvate metabolism in embrionic sensory ganglia. It was found that the sensory ganglia actively metabolize glucose and pyruvate, and the radioactivity which disappeared from the medium was recovered as CO_2, lactate, free amino acids such as aspartate, glutamate and alanine and amino acids incorporated into proteins.

The addition of NGF to the medium increased the amount of radioactivity incorporated into CO_2 when glucose was the substrate and that incorporated into lactate with both substrates.

ANGELETTI _et al_. (1964). _Biochim. Biophys. Acta_ _90_, 445.

THE METABOLISM OF ETHANOLAMINE LIPIDS IN SUB-CELLULAR FRACTIONS OF RAT BRAIN

G.B. Ansell and Sheila Spanner

Department of Experimental Neuropharmacology,
The Medical School, Birmingham, 15.

Previous studies have shown that, when [^{14}C]ethanolamine (1 μc = 0.2 μmoles) was injected intracerebrally into a rat of 12-14 weeks, there was a considerable incorporation into the brain ethanolamine lipids (ANSELL & SPANNER, 1965, 1966). After one hour 9% of the injected material was present in a lipid-bound form; this amount rose to 40% in twenty four hours. The experiments strongly suggested that this ethanolamine was incorporated by the cytidine pathway. This study has now been extended to the sub-cellular particles of brain which were examined for ethanolamine lipids and for the uptake of labelled ethanolamine *in vivo*.

Sub-cellular fractions of whole brain were obtained by a modification of the method of EICHBERG, WHITTAKER and DAWSON (1964) and the ethanolamine lipids from microsomal, mitochondrial, nuclear, supernatant, large and small myelin and debris fractions investigated. When compared with the whole brain homogenate, the recovery in terms of ethanolamine lipid and radioactivity was 70 - 80%. The concentration of ethanolamine lipid in the two myelin fractions was greater than that in any of the others but phosphatidylethanolamine, the ethanolamine plasmalogen and the saturated ether lipid were present in each fraction. Radioactivity was detected in all three ethanolamine lipids in all the fractions, though the amount varied considerably and the turnover as calculated from the specific radioactivities showed a similar diversity. For all three lipids and throughout the 24 hrs. investigated, the greatest count was in the so-called 'small myelin'. In view of its high ethanolamine lipid content this would not be surprising, but in view of its reported inert nature it was unexpected.

In all fractions, labelling of the diacyl lipid was higher than that of the plasmalogen. Determination of the specific radioactivity after short times, showed that the greatest turnover was in the microsomal fraction. However, within 3 hr, the specific radioactivity of phosphatidylethanolamine in the small myelin was similar to that of the microsomal fraction. The specific radioactivity of the plasmalogen in the mitochondrial fraction was always higher than that in any other fraction.

It is certain that there is a considerable exchange of ethanolamine in all the ethanolamine lipids of the sub-cellular fractions examined, with the possible exception of the 'large myelin' fraction. To what extent this represents synthesis of ethanolamine lipids in cell membranes of the brain *de novo* will be discussed.

ANSELL, G.B. & SPANNER, S. (1965). *Biochem. J.*, *96*, 64P.
ANSELL, G.B. & SPANNER, S. (1966). *Biochem. J.*, *100*, 50P.
EICHBERG, Jun., J., WHITTAKER, V.P. & DAWSON, R.M.C. (1964). *Biochem. J.*, *92*, 91.

NEUROCHEMICAL AND NEUROPHYSIOLOGICAL EVIDENCE FOR THE ROLE OF GLYCINE AS AN INHIBITORY TRANSMITTER IN CAT SPINAL CORD

M.H. Aprison, R. Werman, R.A. Davidoff, R.P. Shank and L.T. Graham, Jr.

The Institute of Psychiatric Research, Indiana University Medical Center, Indianapolis, Indiana.

Hypothetical distributions of the excitatory and inhibitory transmitters in spinal cord can be predicted from anatomical and physiological considerations. Neurochemical data from our laboratory show that the distribution of glycine and GABA in lumbar spinal cord follow the distribution of interneurons. Thus, both compounds are present in high concentration in grey matter and in lower concentration in white matter and roots. By means of extracellular iontophoresis both compounds have been shown to inhibit the firing of neurons. Since the inhibitory transmitter is released from interneurons, destruction of these cells should produce a reduction of its concentration. Anoxic loss of spinal interneurons in cats was accomplished by aortic occlusion for 15 to 60 minutes. Following survival for 11 to 35 days, the seventh lumbar segment was removed, frozen, sectioned and the tissue analyzed for cell loss, glycine, GABA and glutamine levels in specific areas. Accompanying the loss of interneurons as observed in histological sections, there was a significant loss ofglycine in the grey matter. The concentration of glycine correlated with the number of remaining interneurons. GABA and glutamine concentrations showed no significant changes and no correlation with neuron count. Neurophysiological studies utilizing intracellular recording and extracellular iontophoresis have demonstrated that glycine and the transmitter released during stimulation of inhibitory nerves affect the postsynaptic membrane of motoneurons in an identical manner. This was shown by equivalance of equilibrium potentials of glycine and the physiological transmitter process before and after injection of intracellular ions. Preliminary experiments have shown that stimulation of spinal nerves caused a release of glycine and several other compounds which react with 2,4-dinitrofluorobenzene, during superfusion of the spinal cord. The results of these investigations are consistent with the hypothesis that glycine is an inhibitory transmitter in the cat spinal cord.

CHANGES IN THE ACTIVITY OF ENZYME SYSTEMS IN EPILEPTOGENIC AND NON EPILEPTOGENIC LESIONS OF THE HUMAN NEOCORTEX

C. Arseni*, G. Badiu**, Nella Mison-Crighel** and M. Simionescu*

The enzymic activity of human cerebral cortex was investigated in samples excised in the immediate vicinity of (a) posttraumatic scars, (b) different types of brain tumours and of (c) haemorrhagic lesions. Enzymic activity was assayed in brain homogenates, immediately after removal of the cerebral cortex, by spectrophotometric methods. The results were calculated, with reference to the concentration of soluble protein.

The following results were obtained:

L-Malate : NADP oxidoreductase (I.U.B. 1.1.1.40) activity was increased in cases with very frequent epileptic seizures and in those with considerable oedema of long duration, indicative of long-lasting circulatory impairment. In recent illness, regardless of the occurrence of epileptic seizures, no change in activity of this enzyme was found.

L-Glutamate : NAD oxidoreductase (I.U.B. 1.4.1.2.) showed an altered activity, increased or decreased in some patients; the activity of L-Lactate : NAD oxidoreductase (I.U.B. 1.1.1.27), succinate (acceptor) oxidoreductase (1.3.99.1) and L_s-isocitrate : NAD oxidoreductase (1.1.1.41) was fractionally increased; no correlation of these enzymic changes with the type of lesion, epilepsy, age of the patient or duration of illness could be established.

L-Aspartate 2-oxoglutarate aminotransferase (2.6.1.1.) was decreased only in cerebral lesions of different type (intracerebral tumours, haemorrhagic brain, scars) independent of the occurrence of convulsions. L-Alanine-2-oxoglutarate aminotransferase (2.6.1.2) activity was lower in all cases of epilepsy but not in other conditions. No relation to the type of lesion could be established. L-Alanine-oxoglutarate aminotransferase activity could be correlated with changes in neocortical excitability.

These preliminary results indicate a metabolic change in cortical tissue taken from the immediate vicinity of brain tumours of different types: they suggest that such tissue cannot be considered normal. Our findings indicate changes of enzymic pattern in cerebral lesions and suggest that there are metabolic disturbances also in the peri-tumoural brain tissue.

* Neurosurgical clinic Bucharest

** Neurochemical laboratory of the Neurological Institute of the R.S.R. Academy.

THE ROLE OF THIAMINE IN THE BIOCHEMISTRY OF NERVOUS TISSUES

A.I. Balakleevsky

Medical Institute, Grodno, USSR

It has been shown in recent years that the role of thiamine in the nervous system is not limited to its function as a coenzyme. The vitamin participates (as thiamine pyrophosphate) in the formation of acetyl-CoA and exercises in this way an influence on the synthesis of acetylcholine; there is also a possibility of the formation of acetylthiamine, and the possibility of an effect of the vitamin upon the activity of cholinesterase and cholinergic receptors.

The transformations of thiamine and its derivatives of thiol-disulphide type, and the participation of thiamine pyrophosphate in the pentosephosphate cycle, are closely connected with the metabolism of thiol substances, the oxidation-reduction transformations of catecholamines and the metabolism of other neurohormones.

The diversity of antagonistic relations between the thiamine derivatives and vitamin B_6 is manifested by changes in the glutamate—GABA, tryptophan—serotonin, histidine—histamine and tyrosine—noradrenaline systems.

We have shown that the hypotensive action of thiamine injected into the circulatory system may be connected as with a direct influence of thiamine on the neuro-receptor system, and with a rapid diminution of the level of serotonin and of the activity of cholinesterase in the blood. Injection of large amounts of thiamine into an organism results in an increase in the quantity of noradrenaline in the brain, of potassium in the blood and adrenals, of sodium in the rat heart and a decrease in the amount of acetylcholine and potassium in the somatic nerves of the pigeon.

The thiamine-antimetabolite pyrithiamine, which is known to influence the nervous system similarly to another thiamine-antimetabolite (oxy-thiamine), increases the quantity of potassium, noradrenaline and 5-hydroxyindoles, and activates serotonin monoamine oxidase in nervous and some other tissues.

These data suggest that the action of thiamine and some of its derivatives and antimetabolites may be explained by the interference of these substances with the metabolism and balance of a number of important neurohormones and electrolytes in the tissues.

EFFECTS OF NEONATAL THYROID DEFICIENCY ON THE DEVELOPING BRAIN

R. Balázs, S. Kovács, P. Teichgräber, W. Cocks and J.T. Eayrs

Medical Research Council Neuropsychiatric Research Unit, Carshalton, Surrey and Institute of Psychiatry, Maudsley Hospital, Denmark Hill, London.

Rats were thyroidectomized at birth by the intraperitoneal administration of 120-200 μC ^{131}I, and their brains were studied after 1-5 weeks. Brain weight was reduced by about 15%, but the DNA content per gram fresh weight was increased by about 10%. The protein and RNA contents, when expressed on a DNA basis were reduced. These results suggest that neonatal thyroid deficiency affects the size of the cells rather than their total number: they are thus consistent with previous histological observations on the growth of the cerebral cortex (see EAYRS, 1960).

In order to determine whether or not the reduction in cell size affected all the cell constituents, the activities of the following enzymes, which served as markers for sub-cellular fractions in homogenates of cerebral cortex, were estimated: (i) lactate dehydrogenase (LDH) for the supernatant; (ii) glutamate dehydrogenase (GDH) for the mitochondrial, and (iii) glutamate decarboxylase (GAD) for the synaptosome fractions. When the enzymic activities were expressed on a fresh weight basis, a significant reduction compared to control values was observed only in the case of glutamate decarboxylase (-15%): when the calculations were based on DNA content all activities were reduced. These observations agree with histological findings, indicating that thyroid deficiency during the maturation of the brain reduces synaptic interaction between neurones (EAYRS, 1960). The absence of significant effect on the activity of GDH must be contrasted with that on another mitochondrial enzyme, succinate dehydrogenase (SDH), shown by HAMBURGH & FLEXNER (1957) to be irreversibly reduced in the brains of cretinous rats. The different effects of thyroid deficiency on the activities of GDH and SDH, together with differences in the sedimentation properties of the particles containing the enzymes, suggests that their distribution in the synaptosomal mitochondria differs from that in other mitochondria in the tissue.

As a preliminary to the investigation of nucleic acid metabolism in the brains of thyroid deficient rats the conversion of (6-^{14}C)orotic acid into pyrimidine nucleotides was studied. Pyrimidine metabolism appeared to be unaffected by neonatal thyroidectomy. The incorporation of protein and nucleic acid precursors in the subcellular fractions of brain at different developmental stages in hypothyroid and in control animals was investigated.

EAYRS, J.T. (1960). Brit. med. Bull., 16, 122. HAMBURGH, M. & FLEXNER, L.B. (1957). J. Neurochem., 1, 279.

METABOLISM OF GABA IN CEREBRAL CORTEX

R. Balázs, Y. Machiyama and D. Richter

Medical Research Council Neuropsychiatric Research Unit, Carshalton, Surrey.

There is good evidence that GABA participates in the utilization of glucose in the brain *in vivo* and in brain slices *in vitro* (ROBERTS, 1962). Certain amino acids, including GABA and glutamate, have also been shown to affect the rate of firing of neurons in the CNS (CURTIS & WATKINS, 1960). In the present investigation an attempt was made to determine (1) the quantitative contribution of GABA oxidation to the total flux through the tricarboxylic acid cycle and (2) the effect of potassium stimulation on the metabolism of GABA.

Guinea-pig brain cortex slices were labelled with [1-^{14}C]GABA, which accumulated in the tissue at a high rate: there was no significant re-entry of ^{14}C from labelled glutamate into the GABA pool. The time-course of incorporation of ^{14}C into different amino acids associated with the cycle was determined and the results were simulated on an analogue computer to permit the calculation of the reaction rates. We are indebted to Mr. J.B. Hammond, NIMR, and Miss T. Julian for the calculations. The fractional rate of GABA utilization was found to be 2 per cent/min. representing about 8 per cent of the total flux in the cycle.

Potassium stimulation with 66m-equiv K^+/l resulted in the following effects on GABA metabolism: (1) About a third of the GABA content of the slices was rapidly released on the addition of K^+ ions. The loss of other amino acids was negligible and the effect was apparently specific for K^+ ions. (2) GABA was taken up subsequently from the medium, but the rate of accumulation was considerably reduced in comparison with controls. (3) The final content of GABA in slices incubated for 1 hr. in a medium containing 66m-equiv K^+/l was approximately double the final steady-state rate in the controls. (4) Computer simulation and independent mathematical analyses of the data showed that both the fractional utilization rate and the formation rate of GABA were altered in comparison with the control.

Comparison of the results obtained when slices were incubated with labelled GABA and [^{14}C]glucose gave evidence of a metabolic compartmentation of GABA: the compartmentation was also affected by potassium stimulation.

CURTIS, D.R. & WATKINS, J.C. (1960). *J. Neurochem.*, *6*, 117.
ROBERTS, E. (1962). In *Neurochemistry*. Edited by K.A.C. Elliott, I.H. Page & J.H. Quastel, p.636. Ch. C. Thomas, Springfield, Ill.

COMPARTMENTILIZATION OF ACETYLCHOLINE IN BRAIN

F. de Balbian Verster and C.A. Robinson

Department of Pharmacology,
Vanderbilt University School of Medicine,
Nashville, Tennessee

The groups of DEROBERTIS and WHITTAKER have shown that acetylcholine (ACh) is distributed in a number of intracellular compartments. WHITTAKER, using density gradient techniques, has defined five different pools: total, free, bound, labile and stable. In order to study the inter relationships of these pools in the brain a technique based on the sequential isolation of tissue fractions was devised. The total amount in each pool is measured, making it possible to determine the absolute distribution of ACh as well as the relative specific concentration of ACh in the pools with respect to protein.

Rats are decapitated and the cerebral cortex removed (total elapsed time, less than 30 seconds). A 1:10 homogenate in 0.32 M-sucrose containing 5.10^{-5} m-moles physostigmine per ml is prepared in the cold. A portion of this homogenate is brought to pH 3.0 with HCl and after three minutes in a water bath spun at 145,000 g for 10 min., the supernatant contains the total ACh. The remainder of 1:10 homogenate is spun for 60 min. at 145,000 g. The supernatant contains the free ACh, the precipitate, the bound ACh. The precipitate is resuspended in water (containing 5.10^{-5} m-moles/ ml physostigmine) and spun again at 145,000 g for 60 min. The supernatant now contains the labile ACh released from the synaptosomes by the osmotic shock treatment with water, and the precipitate contains the stable ACh associated with the synaptic vesicles. The stable ACh is solubilized by reducing the pH to 3.0 with HCl and heating for 10 min. on a water bath. After neutralizing the acidified fractions, they are analysed for ACh by the rat blood pressure response, using the usual precautions.

Under these conditions, the total ACh concentration is 16.5 ± 2.2 (mean $\pm$ S.E.)mμmoles/g wet weight cortex and the free level is $6.0 \pm).5$ mμmoles/g. The bound ACh (calculated from the sum of the labile and stable) is 14.3 mμmoles/g while the labile is 4.4 ± 0.5 mμmoles/g and the stable is 9.9 ± 1.1 mμmoles/g.

The effect of various compounds on the compartmentilization of ACh will be discussed.

Supported by a grant from the United Health Fund and the National Institute of Health (NB06835).

THE INFLUENCE OF POTASSIUM AND GLUCOSE ON GLUTAMINE SYNTHESIS IN RAT CEREBRAL CORTEX SLICES

M. Balliano, I. Masi and F. Pocchiari

Laboratori di Chimica Biologica,
Istituto Superiore di Sanità, Roma, Italy.

It is well known that in rat cerebral cortex slices potassium increase the oxygen uptake (ASHFORD & DIXON, 1935; DICKENS & GREVILLE, 1935) and the incorporation of radioactivity from labelled glucose into glutamine (CHAIN *et al*. 1960). Furthermore it has been shown that in a potassium medium glucose stimulates the formation of glutamine from glutamate (KREBS, 1935) and increases the amount of intracellular glutamate (SELLINGER *et al*., 1962).

Experiments were carried out to find out if the effects of glucose on glutamate metabolism are, or are not, connected with the presence of potassium in the incubation medium.

The results obtained show that potassium increases the conversion of glutamate to glutamine only if glucose is present and glucose, in turn, in the absence of potassium does not stimulate the production of glutamine from glutamate, although the accumulation of intracellular glutamate is enhanced.

It appears therefore that in rat cerebral cortex slices both potassium and glucose are necessary for stimulating glutamine synthesis and this effect is not due to the intracellular concentration of glutamate.

ASHFORD & DIXON (1935). *Biochem. J.*, *29*, 157. CHAIN *et al*. (1960). *Proc. Roy. Soc. B.*, *152*, 283. DICKENS & GREVILLE (1935). *Biochem. J.*, *29*, 1168. KREBS (1935). *Biochem. J.*, *29*, 1951. SELLINGER *et al*. (1962). *Proc. Roy. Soc. B.*, *156*, 148.

MYELINOGENESIS IN RAT BRAIN

N.L. Banik and A.N. Davison

Biochemistry Department, Charing Cross Hospital
Medical School, London, W.C.2.

Myelin isolated by centrifugal procedures from 10-20 day old rat brains has been found to differ in composition from that of adult animals. Much less cerebroside and rather more phospholipid is found in this 'early' myelin and there are changes in its fatty acid composition. The relative deficiency in cerebroside, the high phospholipid and the predominance of short chain fatty acids (16:0) show similarities to the composition of many cellular membranes (ASHWORTH & GREEN, 1966) including probably glial membrane. Thus analysis of enriched glial cell fractions from rat brain and analyses of oligodendroglioma and astrocytoma samples (DAVISON, CUZNER, BANIK & OXBERRY, 1966) show smaller amounts of cerebroside than is present in myelin and a higher proportion of phospholipid. It is suggested that the first stage in myelinogenesis may be the deposition of glial plasma membrane followed by the later insertion of cerebroside to form the mature structure. Newly formed myelin has a different appearance, in tissue culture and in electron micrographs, compared to adult myelin and this may be correlated with the chemical findings. Attempts have been made to delay myelination by administration of triparanol to developing rats. These experiments have led to the appearance of desmosterol in isolated myelin and a relative deficiency of cerebroside was also found (BANIK & DAVISON, 1967). Following discontinuation of the drug the myelin composition slowly returned to normal. These findings and the results from isotopic experiments on myelinating rats will be discussed in relation to the present concept of myelinogenesis.

This work has been generously supported by the Multiple Sclerosis Society.

ASHWORTH, L.A.E. & GREEN, C. (1966). Science, 151, 210.
DAVISON, A.N., CUZNER, M.L., BANIK, N.L. & OXBERRY, J.M. (1966). Nature (Lond.), 212, 1373. BANIK, N.L. & DAVISON, A.N. (1967). J. Neurochem., 14,

COMPARATIVE EFFECTS OF ACETOXYCYCLOHEXIMIDE AND PUROMYCIN ON LEARNING AND MEMORY IN MICE

Samuel H. Barondes and Harry D. Cohen

Departments of Psychiatry and Molecular Biology,
Albert Einstein College of Medicine, Bronx, N.Y.

It has been proposed that memory is based on the synthesis of protein which functions to facilitate synaptic connections. The validity of this proposal is being examined by several groups of investigators by studying the effects of inhibitors of protein synthesis on learning and memory. We have studied the effect, in the mouse, of several inhibitors of brain protein synthesis including puromycin and acetoxycycloheximide, on learning and on memory at various times after learning.

The inhibitors were generally administered intracerebrally in the temporal region of the brain by the method of FLEXNER _et al_. They were given a number of hours before training to allow for intracerebral diffusion. Maximum inhibition was usually achieved 4-5 hours after injection and maintained for the ensuing 3-4 hours. Acetoxycycloheximide (60 micrograms at each temporal site) inhibited 99% of C^{14}-valine incorporation into cerebral protein in the interval between 5 and 8 hours after administration of the inhibitor. Puromycin was somewhat less effective.

Neither puromycin nor acetoxycycloheximide, administered 5 hours before training, influenced the rate of shock-motivated acquisition of the escape-solution of a one-choice T-maze when training and retesting were conducted to the conventional criterion of 9 out of 10 correct responses. Both drugs influenced retention, however. With puromycin there was normal retention for 15 minutes after training, but 3 hours after training there was marked loss of retention which was sustained thereafter. Acetoxycycloheximide did not impair retention 3 hours after training and, when mice were trained to the usual criterion, there was only slight impairment thereafter. The effect of acetoxycycloheximide on retention was much more pronounced if the mice were "undertrained", i.e. trained to a criterion of 3 out of 4 correct responses. In this situation they had normal retention 3 hours after training, but markedly inpaired retention 6 hours after training and thereafter.

The data suggest that cerebral protein synthesis is not required for learning or for memory within several hours after learning. Experiments will be discussed which attempt to distinguish "non-specific" effects of these drugs from a specific inhibition of the synthesis of protein which may mediate the "long term" phase of memory storage.

A CRITICAL EVALUATION OF INTRACEREBRAL INJECTION AND OF ANAESTHESIA IN THE STUDY OF BRAIN METABOLISM *IN VIVO*

Claude F. Baxter, C. Leo Ortiz and David Masuoka

Neurochemistry Labs., V.A. Hospital, Sepulveda; Department of Biochemistry, City of Hope Medical Center, Duarte; and Department of Physiology, University of California School of Medicine, Los Angeles, California.

Interpretation of *in vivo* metabolic studies in which compounds are injected into brain ventricles are complicated by many factors; three of these have been examined.

(1) Non-mixing of an injected metabolite and its products, with all tissue pools of the same compounds. This has been demonstrated in rat brain by a lack of equilibration of ^{14}C from injected γ-aminobutyric acid with total pools of tricarboxylic acid cycle intermediates and glutamic acid.

(2) Needle trauma to capillary networks. The distribution of ^{14}C after the intraventricular injection of γ-amino[2-^{14}C]-butyric acid into unanaesthetized mice (HALEY & MCCORMICK, 1957) was examined by whole body radioautography. Fifteen minutes after injection, levels of radioactivity in liver and kidney correlated with the extent of haemorrhage along the needle track. Radioactivity in urine was almost exclusively in γ-aminobutyric acid. In liver and kidney, labelling was found also in related amino acids.

(3) A significant effect of anaesthetics upon amino acid levels in brain. Stereotaxic injections into rats anaesthetized with pentobarbital resulted in less visible vascular damage, but the metabolic effect of the anaesthetic differed for individual brain areas and was modified by the nutritional state of the animal. Levels of γ-aminobutyric acid were depressed primarily in cortex and caudate nucleus. The degree to which some amino acid levels were lowered by pentobarbitone, halothane and chloralose correlated with the depression of body temperature. In an environment at 22°C, pentobarbitone (45 mg/kg) lowered rectal temperature of rats by 9°C. When normal body temperature was maintained by external heating, glutamic acid in the cerebral cortex remained close to control levels, while levels of γ-aminobutyric acid were depressed. These effects, as well as those upon other amino acids, demonstrated that anaesthetics alter cerebral metabolism *in vivo*, not only by a direct effect upon enzyme systems, but also through a concomitant drop in body temperature.

This investigation was supported by Grant No. NB 03743 from the National Institute for Neurological Diseases and Blindness, USPHS.

HALEY, T.J. & MCCORMICK, W.G. (1957). *Brit. J. Pharmacol.*, **12**, 12.

STUDIES ON THE METABOLISM OF N-ACETYL-L-ASPARTIC-ACID IN PERIPHERAL TISSUES OF THE RAT*

Myron Benuck[1] and Amedeo F. D'Adamo

The Saul R. Korey Department of Neurology and Department of Biochemistry, Albert Einstein College of Medicine, Bronx, N.Y.

A physiological role for N-acetyl-L-aspartic acid as a member of an acetyl transport system for lipogenesis in the brain has been suggested (D'ADAMO & YATSU, 1966). Although uniquely concentrated in the CNS of mammals, small amounts are found in peripheral tissues. It was of interest to determine whether this compound serves as an acetyl source for lipogenesis in peripheral tissues.

Radioactive incorporation into CO_2 and fatty acids from N[1-^{14}C]acetyl-L-aspartic-acid by rat tissue slices was determined. The ratio of fatty acid labelling to that of CO_2 was 6.3 for lactating mammary gland, 0.009 for heart, 0.02 for liver, and 0.02 for kidney. Similar experiments showed wide variation in transport of N-acetyl-L-aspartic-acid into these tissues. Of the added precursor, 32% was taken up by mammary gland slices of which approximately 50% was converted into fatty acids. Since this high conversion is similar to that found in myelinating rat brain, incubations with labelled glucose and acetate were performed to determine whether N-acetyl-L-aspartic-acid synthesis occurs in mammary gland. After homogenization in chloroform-methanol, the aqueous extract was dried and carrier added. The purification sequence was: column chromatography, two paper chromatograms, one acidic and one basic, and a final Dowex column. Under the conditions employed, only in the glucose experiment was significant radioactivity found in the N-acetyl-L-aspartic-acid fraction. Addition of carrier and recrystallization as the anhydride removed all remaining radioactivity.

These results indicate that, although an active coupling mechanism for conversion of the acetyl amino acid to fatty acids exists in mammary gland, any N-acetyl-L-aspartic-acid required by this tissue is supplied by the brain.

*Supported by U.S. Public Health Service grants NB-03356 and NB-04903.

[1]Postdoctoral Fellow, U.S. Public Health Service grant 5TI-MN-6418.

D'ADAMO, A.F. & YATSU, F.M. (1966). J. Neurochem., 13, 961.

GLYCOLYSIS AND VISUAL CELL FUNCTION

E. De Berardinis and O. Tieri

University Eye Clinic, Napoli, Italy

The function and viability of the visual cell depend on glycolysis. Inhibition of the glycolytic activity extinguishes the electrical response to light stimulation and damages irreversibly the retinal neuroepithelium. Blocking a single glycolytic enzyme is enough to do that.

It has been shown that there is a parallelism between enzyme inhibition and ERG extinction. It has also been demonstrated that there is a recovery of the glycolytic activity on adding to the system the inhibited enzyme.

Finally, it has been shown that the glycolytic block *in vivo* always coincides with a retinotoxic action, provided that the metabolic damage really affects the visual cells.

THE EXISTENCE OF TWO PATHWAYS FROM ACETYL CoA TO GLUTAMATE IN THE CENTRAL NERVOUS SYSTEM

C. van den Berg*, Lj. Kržalic, P. Mela

New York State Psychiatric Institute, New York

[1-^{14}C]Glucose, [2-^{14}C]glucose and [1-^{14}C]acetate, [2-^{14}C]acetate were injected intraperitoneally in mice. We determined the specific activities of glutamate, glutamine and aspartate in the brain at 5, 15, 30 and 45 minutes after the injection.

The specific activity of glutamine was always higher than the specific activity of glutamate when acetate was used as a precursor. This was not the case with glucose as precursor.

The ratio of incorporation of the C-1 to C-2 of glucose and the C-2 to C-1 of acetate in glutamate and aspartate increased with increasing time after injection. This increase was less for glutamine, especially when acetate was used as precursor.

After acetate the amount of ^{14}C in the C-1 of glutamine was less than in the C-1 of glutamate. A model consisting of two types of tricarboxylic acid cycles will be proposed to explain these and relevant data from the literature. Preliminary results on the intracellular localization of three enzyme involved in these reactions suggest that a morphological basis exists for these two types of cycles.

*Present address: Central Institute for Brain Research, Amsterdam.

REGIONAL METABOLISM OF [^{14}C]LEUCINE IN BRAIN (CAT): VENTRICULAR AND INTRAVENOUS ADMINISTRATION.

Soll Berl and Tamas Frigyesi

Columbia University, College of Physicians and Surgeons, New York, N.Y.

Studies on [U-^{14}C]leucine metabolism in vivo in the cat brain was attempted by administration of radioisotope through an anatomically verified stereotaxically placed cannula into the 3rd ventricle. After 20 minutes the brain was dissected into right and left areas of cortex, cerebellum, pons, medulla, thalamus, caudate nucleus and hippocampus. The distribution of radioactivity in the different brain areas showed pronounced bilateral variation. Conversion of leucine to glutamate, glutamine and aspartate was very low and the values for the half life of the TCA-insoluble protein was very long. These data suggest that the ventricular ependymal lining serves as a retaining barrier for the underlying tissue.

Following intravenous administration of labelled leucine, the radioactivity was rather evenly distributed bilaterally in all areas. The percentage of total radioactivity in glutamic acid ranged approximately from 10% (brain stem) to 37% (cerebellum), in glutamine from 9% (cortex, caudate nucleus) to 13% (brain stem), and in aspartic acid from 1.5% (hippocampus) to 6.3% (cerebellum). The half life of the protein ranged from approximately 4 days (cerebellum) to 9 days (brain stem).

The specific activities (counts/min/μmole) of the free tissue and plasma leucine were approximately equal at the termination of the experiment, indicating attainment of a steady state. This was not true for glutamic acid, glutamine and aspartic acid; the data indicate formation of these amino acids in the brain. The relative specific activity of the glutamine (glutamic acid = 1) was significantly greater than 1 in all areas except cerebellum and caudate nucleus. However, the cerebellum and caudate did reveal glutamine R.S.A. values greater than 1 following intravenous administration of [1-^{14}C]acetate. The data support compartmentation of the glutamate-glutamine system in all these brain areas. These biochemical properties of the striatum and cerebellum are probably based on their differential structural characteristics.

INVESTIGATIONS ON THE SIGNIFICANCE OF CEREBROSPINAL FLUID CREATINE

H.H. Berlet and C. Quarcoo

Neurologische Klinik und Poliklinik der Universität,
Department of Neurochemistry, Göttingen, Germany.

The method of TANZER & GILVARG (1959) for the enzymic determination of creatine was adapted for the determination of serum creatine by eliminating pyruvate and other substances known to interfere with this assay. The procedure was used to evaluate the relationship between creatine in serum and cerebrospinal fluid (CSF) in man.

Creatine was determined in samples of blood and CSF obtained simultaneously from the subjects under investigation. In adults a striking difference was found to exist between creatine values in serum and CSF: CSF concentrations were significantly higher than those found in serum, e.g. by a factor of 10 in some cases. No such gradient exists in children with physiologically elevated creatine levels and in patients with high serum creatine levels due to various disease states. The observation of this gradient is somewhat at variance with findings reported in the literature and, therefore, the uptake of orally administered creatine into the CSF spaces was also investigated.

The results of these studies indicate the presence of a barrier between blood and CSF spaces for creatine in man, apparently resulting in an accumulation of creatine in the CSF, so that CSF creatine levels are high in comparison with serum creatine levels. The implications of this observation will be briefly considered.

TANZER, M.L. & GILVARG, C. (1959). J. biol. Chem., 234, 3201.

STUDIES ON THE NATURE OF THE ENDOGENOUS SUBSTRATE IN THE BIOSYNTHESIS OF CEREBROSIDE SULPHATE IN BRAIN

V.R. Bhandari and B.K. Bachhawat

Department of Neurology, Neurosurgery and Neurochemistry,
Christian Medical College Hospital,
Vellore, S. India

Earlier studes from this laboratory have demonstrated the *in vitro* formation of cerebroside sulphate ester (CSE) by the transfer of sulphate from 3'-phosphoadenosine 5'-phosphosulphate (PAPS) to an endogenous acceptor present in a crude extract obtained from brain. Galactose oxidase treatment and tryptic digestion of the extract indicated the protein-bound galactocerebroside nature of this endogenous substrate. These observations led us to make further investigations into the nature of this endogenous substrate.

A preliminary lipid analysis using florosil and DEAE-cellulose column chromatography and thin-layer chromatography revealed the presence of both CSE and cerebrosides in the crude extract prepared from the white matter of the sheep brain. A partially purified fraction was obtained by subjecting the crude extract to protamine sulphate and ammonium sulphate fractionations. This fraction was found to be free from CSE and contained 70 mμmoles of cerebroside/100 mg protein: it could convert approximately 10% of the endogenous cerebrosides into CSE. Gel filtration on Sephadex G-75 and G-100 columns could not separate the substrate from the enzyme since both were eluted in the void volume. However, preliminary studies using Sephadex G-200 columns indicated a partial separation of the two activities.

In another experiment, the fraction was prepared as outlined above from the brain of young rats which had earlier been injected with radioactive sulphate. No radioactive CSE could be detected in this fraction, thereby, ruling out the possibility of a sulphate exchange reaction between CSE and PAPS.

These findings indicate that the endogenous substrate is a protein-bound cerebroside and there is a direct transfer of sulphate from PAPS to this substrate during the formation of CSE.

$$\text{Protein-cerebroside} \xrightarrow[\text{PAPS}]{\text{Sulphokinase}} \text{Protein} + \text{CSE} + \text{PAP}$$

The nature and the specificity of the protein moiety still remain to be determined.

The work was supported by Grant No. 614-7 from the National Multiple Sclerosis Society, Grant No. NIH-IND-6X4324 from the National Institute of Health, U.S. Public Health Service and Council of Scientific and Industrial Research (India).

CEREBRAL METABOLISM OF L-[U-^{14}C]GLUTAMINE AND OF L-[U-^{14}C]GLUTAMIC ACID IN CHRONIC MENTAL DISEASE AND IN THERAPEUTIC INSULIN COMA

G. Bianchi Porro, A.T. Maiolo, P. Della Porta,
E. Rossella and E. Polli

Istituto di Semeiotica Medica dell'Università di Milano and
Istituti Psichiatrici della Provincia di Milano

L-[U-^{14}C]glutamine and L-[U-^{14}C]glutamic acid were used as substrates for determining cerebral metabolism in humans *in vivo*, before and during insulin coma. An 80 μc i.v. injection of L-[U-^{14}C]glutamine was given to four patients and an 80 μc i.v. injection of L-[U-^{14}C]glutamic acid was given to three patients, all of whom were suffering from schizophrenia.

(1) Under basal conditions the following results were observed:

(a) a late production by the brain of $^{14}CO_2$ was found (4 - 10 minutes after the injection)

(b) labelled GABA was found in the cerebral venous blood in one case after L-[U-^{14}C]glutamine administration and in two cases after L-[U-^{14}C]glutamic acid administration

(c) neither labelled pyruvic nor α-oxoglutaric acids were ever observed

Such results would support the hypothesis that passage through the blood-brain barrier of both these amino acids takes place and they confirm the possibility that their cerebral metabolism involves the GABA shunt.

(2) During hypoglycemic coma the data observed seem to confirm a cerebral uptake of the injected L-[U-^{14}C]glutamine; in fact there was found

(a) an early cerebral production of $^{14}CO_2$ and

(b) a release in the cerebral venous blood of labelled GABA.

(3) Neither a production of $^{14}CO_2$ nor the presence of labelled GABA was observed after L-[U-^{14}C]glutamic acid administration during insulin coma.

The data suggest that there was a more rapid, but not a quantitatively increased uptake of the L-[U-^{14}C]glutamine from the blood during insulin coma, whereas the brain would not metabolize the glutamic acid.

A significant increase, during hypoglycemic coma, of the cerebral blood flow was confirmed, while the cerebral metabolic rate remained within normal limits.

ACTIVATION AND SOLUBILIZATION OF PARTICLE-BOUND BRAIN HEXOKINASE

D. Biesold and P. Teichgräber

Department of Neurochemistry, Karl Marx University, Leipzig, G.D.R.

Studies of enzyme distribution have shown that most of the hexokinase activity is located in the mitochondrial fraction (BIESOLD & CANZLER, 1965). Experimental data will be discussed, in which activation and solubilization of particle-bound hexokinase in the brain mitochondrial fraction were studied. This fraction was isolated by the method of BALÁZS *et al.* (1963); hexokinase activity was estimated according to BENNETT *et al.* (1962) with glucose-6-phosphate dehydrogenase.

Incubation of brain mitochondrial preparations for 20 min. at 0° with salts and detergents respectively, resulted in a maximal $2\frac{1}{2}$-fold increase of hexokinase activity. Incubation at pH 7 in media containing 0.9 M-NaCl increased hexokinase activity to 180% of the original activity. After centrifugation at 100,000 g for 1 hr., 130% of this activity was found in the supernatant only. Activation was observed not only in the presence of NaCl, but also with Na_2SO_4, Na-acetate, Na-citrate, KCl, and $(NH_4)_2SO_4$: a dependence on the anions and cations used was not seen. Activation was also observed with detergents. Of the detergents studied Triton X-100 showed under our experimental conditions (incubation for 20 min. at 0° with 0.5% final concentration of the detergent) the greatest activation. However, incubation of mitochondria in media containing 0.5% Triton X-100 plus 0.9 M-NaCl increased the activity above that obtained with Triton X-100 or NaCl alone. The activity observed in the presence of Triton X-100 plus NaCl was released in the supernatant after centrifugation at 100,000 g for 1 hr. These results show that the latent form of hexokinase can also be solubilized.

Whereas extraction with salts solubilized only 10% of the mitochondrial proteins, treatment with salt plus Triton X-100 solubilized nearly 90%. Therefore under the latter conditions the increase in specific hexokinase activity of the supernatant was only 2-fold, whereas after treatment with salt alone it was 10-fold. From these results it can be concluded that part of the particle-bound hexokinase activity is latent. From the solubilization experiments one must conclude that hexokinase is bound to different structures.

BALÁZS, R., BIESOLD, D. & MAGYAR, K. (1963). J. Neurochem., 10, 685. BENNETT, E.L., DRORI, J.B., KRECH, D., ROSENZWEIG, M.R. & ABRAHAM, S. (1962). J. biol. Chem., 237, 1758. BIESOLD, D. & CANZLER, E. (1965). Wiss. Z. Karl-Marx-Univ. Leipzig, Math.-Naturw. Reihe 14, 691.

PIGEON BRAIN PROTEINS: STRUCTURE, BIOSYNTHESIS AND STATE IN LEARNING

Samuel Bogoch

Boston University School of Medicine, Boston.

Quantitative methods have been recently described by us for the extraction of brain proteins, and their separation by column chromatography, disc gel electrophoresis and carbohydrate analysis (QUAMINA & BOGOCH, 1965; BOGOCH *et al.*, 1966). These methods have now been applied to the study of (1) human brain glial tumours *in situ*, (2) human glial tumours transplanted to mouse subcutaneous loci, and (3) tissue cultures of human glial tumours. Components of brain protein fractions 10B, 11A, 12 and 13 have been shown to be present in each of the above glial cell growths, and to be predominantly associated with fractionated glial cell membranes.

Male white Carneaux pigeons were studied in control groups and in two experimental groups: one sacrificed immediately after, and another many months after, operant conditioning training with non-aversive, non-stressful, light and sound stimuli. Significant quantitative changes occurred in protein groups 2, 10B and 11A which could be related to training. Incorporation of radioactivity from [U-^{14}C]glucose given *in vivo* was high in protein fractions 10B, 11A, 12 and 13, which were also shown to be the four fractions with the greatest content of bound carbohydrate residues. Hydrolysis of these protein fractions was followed by thin layer chromatography of the liberated carbohydrate and amino acid residues. Quantitative studies of these constituents and their radioactive labelling will be discussed.

BOGOCH *et al.* (1966). In *Inborn Disorders of Sphingolipid Metabolism.* Edited by S.M. Aronson & B.W. Volk. p.273. Pergamon, Oxford. QUAMINA, A. & BOGOCH, S. (1965). In Protides of the Biological Fluids. Edited by H. Peeters. p.211. Elsevier, Amsterdam.

STUDIES ON FREE NUCLEOTIDES AND RIBONUCLEIC ACIDS OF THE RAT BRAIN UNDER VARIOUS FUNCTIONAL CONDITIONS

Vincenzo Bonavita

Department of Neurology, University of Palermo,
Palermo, Italy

Free nucleotides of the rat brain have been separated chromatographically by the procedure of MANDEL and HARTH (1961) with minor modifications. Quantitative measurements of mono-, di-, and tri-phosphorylated nucleotides were made after intraperitoneal injection of uracil, uridine and uridylic acid (10 ml/kg body weight of a solution 7.5×10^{-2} M). A significant increase of uridine 5-phosphate (UMP), adenosine 5-diphosphate and adenosine 5-triphosphate was found following the injection of the nucleoside and the nucleotide, while no changes were seen after uracil injection.

When tritiated uridine [5,6-^{3}H]uridine was given by intraperitoneal or subarachnoid injection, UMP, uridine 5-diphosphoglucose, uridine 5-diphosphate N-acetylglucosamine, uridine 5-triphosphate, cytidine 5-phosphate (CMP), and cytidine 5-triphosphate were labelled. It was shown that the turnover of uracil derivatives is greater than that of CMP, and preliminary evidence in favour of the existence of separate pools of UMP in the mature rat brain were obtained.

Ribonucleic acids were extracted by phenol in the presence and absence of sodium dodecylsulphate, and separated on a sucrose density gradient (5-20%) by centrifugation at 20,000 r.p.m. for 15 hr in a Spinco ultracentrifuge (model L). The 4-S fraction, a high molecular weight component, and components in the region of the 28-S and 18-S fractions showed different rates of incorporation of [5,6-^{3}H]uridine given by subarachnoid injections.

Turnover rates of all the free nucleotides and RNA fractions mentioned above were measured also in convulsant and anaesthetized animals. The physiological significance of changes in turnover rates under these conditions will be discussed.

MANDEL & HARTH (1961). J. Neurochem., 8, 116.

COMPARATIVE STUDY OF NUCLEIC ACIDS OBTAINED FROM RABBIT LIVER AND BRAIN

T. Borkowski

Department of Physiological Chemistry, Medical School, Lublin, Poland

Mitochondria were prepared from rabbit whole brain or liver after differential centrifugation in isotonic sucrose solution.

We found by electron microscopic examination a high morphological purity of the mitochondrial pellets. Phenol was used under different conditions of temperature, pH and NaCl concentration for extraction of nucleic acids from the mitochondria. The nucleic acids were chromatographed on methylated albumin-Kieselguhr columns and different fractions were separated. The nucleotide composition of each nucleic acid fraction was analysed. The chromatographic pictures of the two types of mitochondrial nucleic acids were different.

We found a small quantity of t-RNA and high concentration of r-RNA in liver mitochondria. In each sample of nucleic acid obtained from brain mitochondria we demonstrated three distinct fractions corresponding to t-RNA, r-RNA and DNA.

The base composition of mitochondrial DNA was compared with that of nuclear DNA in liver and brain.

METABOLISM OF NERVE-ENDING PARTICLES

H.F. Bradford

Medical Research Council Metabolic Reactions Research Unit, Imperial College of Science, London, S.W.7.

Suspensions of nerve-ending particles prepared by differential and density-gradient centrifugation in sucrose or Ficol (WHITTAKER, 1959; KUROKAWA, SAKAMOTO & KATO, 1965; ABDEL-LATIF, 1966) provide an opportunity for study of the metabolic behaviour of a cerebral tissue compartment derived specifically from neurons, and, moreover, directly involved in electrical transmission between neurons.

Although the preparative procedures may involve several hours' exposure to temperatures of 0-4° and to low oxygen tensions, these particles display considerable oxidative capacity, absorbing oxygen at rates of 110 ± 28 μmoles O_2/100 mg protein/hr with 10 mM-pyruvate, and 88 ± 13 μmoles O_2/100 mg protein/hr with 10 mM-glucose in Krebs-Ringer media buffered with tris-HCl or glycylglycine at pH 7.4.

Both exogenous [C^{14}]glucose and [C^{14}]pyruvate are transformed by the particles into glycolytic and citric acid cycle intermediates with parallel formation of glutamate, aspartate, alanine and γ-aminobutyric acid. A detailed quantitative comparison of these transformations will be presented.

Although nerve-ending particles possess a potent Na^+-K^+-ATPase (HOSIE, 1965; KUROKAWA *et al.*, 1965; BRADFORD, BROWNLOW & GAMMACK, 1966) they show little ability to accumulate potassium; a limitation in common with preparations of neuronal cell bodies (BRADFORD & ROSE, 1967).

ABDEL-LATIF, A.A. (1966). *Biochem. Biophys. Acta* 121, 403. BRADFORD, H.F., BROWNLOW, E.K. & GAMMACK, D.B. (1966). *J. Neurochem.*, 13, 1283. BRADFORD, H.F. & ROSE, S.P.R. (1967). *J. Neurochem.* In Press. HOSIE, R.J.A. (1965). *Biochem. J.* 96, 404. KUROKAWA, M., SAKAMOTO, T. & KATO, M. (1965). *Biochem. J.*, 97, 833. WHITTAKER, V.P. (1959). *Biochem. J.*, 72, 694.

CORRELATION BETWEEN INDUCED CHANGES IN THE ELECTROCORTICOGRAM AND THE RELEASE OF BIOLOGICALLY ACTIVE SUBSTANCES FROM THE CEREBRAL CORTEX

P.B. Bradley, Gillian M.R. Samuels and Jane E. Shaw

Department of Experimental Neuropharmacology,
The Medical School, Birmingham 15.

Previous studies have shown that various biologically active substances, such as acetylcholine (MITCHELL, 1963) and prostaglandins (RAMWELL & SHAW, 1966), are released from the cerebral cortex of the anaesthetized cat. Although the quantities of prostaglandins released were found to increase during stimulation of a peripheral nerve or of the cortex, it is not known how far these quantities were influenced by the presence of anaesthetic.

The experiments reported here were carried out on unanaesthetized *encéphale isolé* cat preparations in which the electrical activity of the brain was recorded concomitantly with the collection of superfusate. The latter was assayed for catecholamines, 5-HT and prostaglandins.

The amounts of prostaglandins released were higher than those seen in anaesthetized preparations. Furthermore, electrical stimulation of the brain stem reticular formation which resulted in electrocortical arousal, led to an increase in prostaglandin release. Chlorpromazine not only reduced the basal level but also modified the increased release on stimulation.

No detectable quantities of catecholamines or 5-HT were found in the superfusates with the methods employed.

MITCHELL, J.F. (1963). *J. Physiol.*, *165*, 98. RAMWELL, P. & SHAW, J.E. (1966). *Amer. J. Physiol.*, *211*, 125.

CYCLIC NUCLEOTIDE PHOSPHODIESTERASE IN BRAIN - QUANTITATIVE HISTOCHEMISTRY

Bruce McL. Breckenridge and Robert E. Johnston

Department of Pharmacology and Beaumont-May Institute of Neurology, Washington University School of Medicine, St. Louis, Missouri.

Adenosine 3',5'-phosphate (3,5-AMP), which mediates certain actions of catecholamines and of 5-hydroxytryptamine, is inactivated by hydrolysis of the 3'-P bond. The enzyme responsible, cyclic nucleotide phosphodiesterase, occurs at a high level of activity in the brain (DRUMMOND & PERROTT-YEE, 1961). Some concentration of enzyme activity has been noted in synaptosomes (CHEUNG & SALGANICOFF, 1966) and in glial cells (SHANTA *et al*., 1966).

A quantitative histochemical procedure for the measurement of phosphodiesterase activity has been developed and applied to various regions of the rabbit central nervous system. Portions of frozen-dried brain (0.5-2.0 μg) are incubated at 37° in a combined reagent containing imidazole buffer pH 7.6, 2 mM-$MgCl_2$, 4 mM-3,5-AMP and excess snake venom 5'-nucleotidase. The inorganic phosphate released is measured in a second step by means of reactions catalyzed by glycogen phosphorylase, phosphoglucomutase and glucose-6-P dehydrogenase. The TPNH formed as a final product is measured fluorometrically. Typical values for the rate of hydrolysis of 3,5-AMP are (moles/kg dry wt/hr): caudate nucleus 9.4; cerebral cortex 7.3; cerebellum, molecular layer 3.0, granular layer 2.2, white matter 0.2; hypothalamus, ventro-medial nucleus 6.0; arcuate nucleus 3.7, preoptic nucleus 5.6; spinal cord, dorsal columns 0.06, ventral-horn grey 0.2. The wide range of values observed is more characteristic of enzymes associated with specialized functions, such as transmitter metabolism than with energy metabolism.

(Supported by Grant NB-01352 from the National Institute of Neurological Diseases and Blindness).

CHEUNG & SALGANICOFF (1966). *Fed. Proc.*, *25*, 714.
DRUMMOND & PERROTT-YEE (1961). *J. biol. Chem.*, *236*, 1126.
SHANTA *et al*. (1966). *Histochemie*, *7*, 177.

INCORPORATION OF [^{3}H]URIDINE INTO MOUSE BRAIN RNA DURING STRESS

Robert N. Bryan and Eugene L. Bliss

University of Utah College of Medicine, Salt Lake City, Utah

The animals which have been used in most studies attempting to show a relationship between brain RNA metabolism and learning suffered some stress during the training and testing procedures. We thought that the effect of stress on brain RNA metabolism should be investigated. Adult male C3Hf mice were injected intraperitoneally with [5-^{3}H]uridine. Control animals were returned to their cages, and experimental animals were completely immobilized by taping them to wood dowels. Neither group received food or water during the experiment. Animals were killed at various intervals from 0.5 to 72 hr after injection, and radioactivity was estimated in plasma, brain acid-soluble fraction, and brain cytoplasmic and nuclear RNA. Stressed mice incorporated more label into brain nuclear RNA than control mice from 1 to 24 hr after injection. The increased incorporation could not be accounted for by increases in plasma or acid soluble radioactivity. Comparable measurements in liver after 2 hr of stress showed that incorporation into the liver acid soluble fraction decreased 24% and incorporation into liver nuclear RNA decreased 42%. We conclude that during stress RNA synthesis increases in brain and decreases in liver.

Supported by a grant from NIH.

CHANGES IN THE CONTENT OF INDIVIDUAL PHOSPHOLIPIDS IN BLOOD ENTERING AND LEAVING THE BRAIN AND IN THE CEREBROSPINAL FLUID OF THE DOG, AND OF TOTAL PHOSPHOLIPIDS OF RAT BRAIN, DURING VARIOUS FUNCTIONAL STATES

H.Ch. Buniatian and C.G. Karageosian

Institute of Biochemistry of the
Armenian Academy of Sciences, Erevan, U.S.S.R.

The total phospholipids (P) of whole blood were separated by ascending chromatography on paper impregnated with silicic acid into: (1) unidentified acid P (first obtained by us), (2) lysolecithins (LL), (3) monophosphoinosidephosphatides (MP), (4) sphingomyelins (S), (5) lecithins (L), (6) serinephosphatides (SP), (7) ethanolaminephosphatides (EP).

Intraarterial administration of γ-aminobutyric acid (GABA) and adrenaline, and electrocutaneous stimulation (ECS), induced an elimination of P from the brain as shown by arteriovenous (A-V) difference. GABA, and especially ECS, increased the neutral P and reduced the acid P of blood leaving the brain. Adrenaline increased the content of all P fractions except the LL. Such an effect was not seen in blood from peripheral muscle, which suggests a possible functional correlation between the above-mentioned changes and the metabolism of nervous tissues.

A parallel combined study of the blood circulation rate in the common carotid artery and the white matter of the brain made us conclude that the A-V difference in the content of lipid phosphorus, which increased at the 20th min after stimulation, could be ascribed to changes of blood circulation rate, inasmuch as these take place only during the initial period (first 3-4 min).

During insulin-induced hypoglycaemia there was a positive A-V difference in total P which was due mainly to increase of L. During this the lipid phosphorus of the cerebrospinal fluid (CSF) of dogs decreased on account of a reduction of diphosphoinosidephosphatides (DP), MP, S and L, which was not compensated by a five-fold increase of SP.EP and polyglycerophosphatides (PP) of the CSF were not changed. The content of total lipid phosphorus of rat brain did not change following fasting (4-6 days), since the decrease of MP, S and PP were balanced by an increase of SP and EP, while DP and L did not change. During experimental alloxan diabetes in rats the level of total lipid phosphorus was reduced, which according to preliminary experiments was due to a decrease of L.

ON THE MECHANISM OF FORMATION OF AMMONIUM FROM AMINO ACIDS IN BRAIN

H.Ch. Buniatian and S.G. Movcessian

Institute of Biochemistry of the
Academy of Sciences of the Armenian SSR, Yerevan, USSR.

The mechanism of the formation of ammonium from amino acids remains obscure. Our experiments have shown that the addition of nicotinamide-adenine-dinucleotides (NAD, NADH) to brain tissue preparations, and to the mitochondrial fraction of the rat brain in particular, resulted in a considerable liberation of ammonium. Closer examination of this phenomenon indicated that the ammonium is formed from NAD at the expense of its adenine moiety; thus NAD is converted to deamino-NAD (D-NAD), the adenylic acid residue being replaced by inosinic acid (IMP). Bearing in mind the reamination of IMP to adenylic acid by aspartate (ASP) through the formation of adenylosuccinate we studied the formation of ammonium following the addition of D-NAD and ASP. The results obtained showed that in mitochondrial fractions of brain tissue the combined addition of D-NAD and ASP brought about a considerable liberation of ammonium. This process was more pronounced when ASP was also added. In this system glutamate and γ-aminobutyrate were less effective than ASP, and IMP much less than D-NAD. The deamination of NAD occurred also in the mitochondrial fraction of rabbit liver. A marked increase of ammonium concentration was observed following the addition of D-NAD and ASP. The data obtained allow us to propose the following pathway for the formation of ammonium from amino acids:

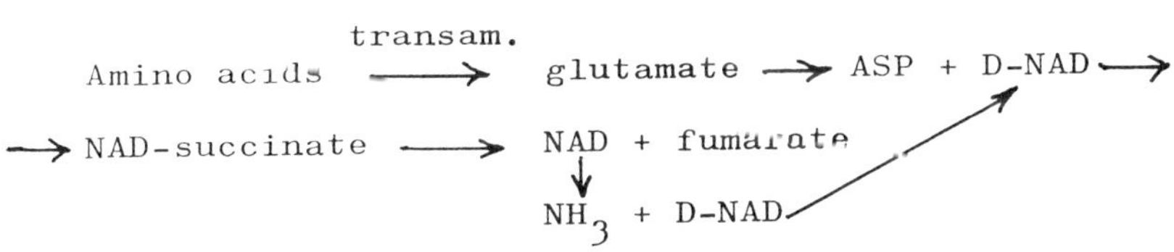

Our experiments indicate that NAD is deaminated by mitochondrial fractions as actively as AMP, but the deamination of these substances is effected by different deaminases. FAD is also deaminated by brain mitochondrial fractions. The deamination of NAD is of importance, not only in amino acid metabolism, but in oxidative processes as well.

SIZE DISTRIBUTION OF RNA ISOLATED FROM GOLDFISH BRAIN

Luigi Casola and Bernard W. Agranoff

International Laboratory of Genetics and Biophysics, Naples, Italy, and Mental Health Research Institute, University of Michigan, U.S.A.

The RNA of whole goldfish brain resembles that of mammalian tissues. 28S, 18S and 4S peaks were seen following sucrose gradient (4-20%) centrifugation of phenol-extracted RNA. In order to study RNA synthesis, [5-^{3}H]uridine was injected intracranially into groups of 20 goldfish and the brains were quickly removed at varying times later. Nuclear or cytoplasmic fractions were obtained and treated with hot phenol-sodium dodecyl sulphate. At 2.5, 5 and 10 minutes after injection, most of the radioactivity was in the nuclear fraction and had a broad size distribution from 18S to > 45S, including the characteristic 35S and 45S ribosomal precursor RNA's. Radioactivity appeared at later times in the cytoplasmic fraction, in the 4S to 18S region. At 20 minutes, radioactivity in the 18S peak had increased and labelling appeared in the 28S region. When deoxycholate-treated post-mitochondrial fractions were sedimented in a 15-30% sucrose gradient, the 45S region was labelled 10 minutes following [5-^{3}H]uridine injection, while the 65S region was labelled in addition at 20 minutes. These particles correspond in size to the known ribosomal subunits, and in their labelling sequence to 18S and 28S cytoplasmic RNA.

(Supported by grants from the National Institutes of Health and National Science Foundation).

INCORPORATION DE L'ACIDE [2-^{14}C]MEVALONIQUE DANS LE CHOLESTEROL DU SYSTEME NERVEUX DU RAT ADULTE. LOCALISATION ANATOMIQUE DES SITES DE SYNTHESE.

F. Chevallier, C. Gautheron and L. Petit

Département de Biologie,
Centre d'Etudes Nucléaires de Saclay, 91-Essonne

Le but de cette étude est la détermination par autoradiographie des régions anatomiques de l'encéphale du rat adulte ayant la propriété de synthétiser le cholestérol. Une telle recherche nécessite l'emploi d'un précurseur spécifique de la synthèse du cholestérol.

On a recherché, d'une part, dans quelles conditions l'acide mévalonique répondait à ce critère en mesurant la radioactivité du $^{14}CO_2$ expiré par l'animal et celles du cholestérol, des lipides et de la fraction saponifiable dans le cerveau, la moelle, le foie et le sang. Il est nécessaire de sacrifier l'animal 72 heures environ après l'administration du précurseur pour que la radioactivité du cerveau ne corresponde pratiquement qu'à celle du cholestérol.

On a recherché, d'autre part, quelle était la meilleure voie d'administration du mévalonate en comparant les radioactivités spécifiques du cholestérol des organes cités ci-dessus. Diverses raisons conduisent à écarter l'administration par voie intrapéritoniale. La perfusion continue (4 à 8 heures) dans le liquide céphalorachidien permet d'obtenir des taux d'incorporation tels qu'une étude autoradiographique peut être entreprise. Mais la grande vitesse de diffusion du précurseur du liquide céphalorachidien dans la circulation sanguine mesurée en continu par le $^{14}CO_2$, et la très faible vitesse de diffusion du mévalonate dans le liquide céphalorachidien obligent, pour avoir, par autoradiographie, une image complète des sites de synthèse, à utiliser deux voies d'administration: intracisternale et intraventriculaire.

L'étude autoradiographique est en cours de dépouillement.

ETUDE DES TRANSFERTS DE CHOLESTEROL D'ORIGINE PLASMATIQUE DANS LE SYSTEME NERVEUX CENTRAL DU RAT ADULTE

F. Chevallier and L. Petit

Département de Biologie,
Centre d'Etudes Nucléaires de Saclay, 91-Essonne

Chez le rat adulte, les échanges de cholestérol entre un organe et le plasma constituent le mode essentiel de renouvellement du cholestérol cellulaire. Le cholestérol du système nerveux central ne fait pas exception à cette règle. Mais ce processus est caractérisé, à ce niveau, par deux faits. D'une part, seule une fraction du cholestérol est ainsi renouvelable: 23, 12, 15 pour 100 respectivement pour le cerveau, la moelle épinière, le nerf sciatique. D'autre part, les vitesses des transferts sont lentes. Elles sont telles que les pourcentages cités ci-dessus n'ont pu être déterminés qu'après un an d'ingestion quotidienne de [^{14}C]-cholestérol. On a, par ailleurs, montré qu'au transfert plasma-cerveau est associé un transfert en sens inverse cerveau-plasma. Il s'agit donc bien d'un renouvellement par échange ou remplacement. Ces faits soulignent aussi la relative imperméabilité de la barrière hémato encéphalique.

Ces données acquises, on a voulu préciser si la fraction du cholestérol du cerveau renouvelable par transfert était localisée dans des régions anatomiques particulières. Ce problème a été abordé par ^{14}C-autoradiographie. On a constaté que le noircissement des films était rigoureusement proportionnel à la densité des fibres myélinisées. Pour avoir plus de précisions sur la localisation du cholestérol de transfert au niveau de ces zones à fibres myélinisées, des expériences ont été entreprises par radioautographie en microscopie électronique en utilisant du cholestérol tritié. Ce travail est en cours, en collaboration avec DROZ B.

ACTIVATION OF L-HISTIDINE DECARBOXYLASE BY A PYRIDOXAL-5'-PHOSPHATE DERIVATIVE

William G. Clark, Dorothea Aures and Charles Wiseman

V.A. Hospital, Sepulveda, Calif., and Department of Biological Chemistry, Medical Center, University of California, Los Angeles.

Study of the activation of apo-decarboxylase, *in vitro*, by pyridoxal-5'-phosphate (PALP), is complicated by the reactivity of the carbonyl group of this coenzyme which combines with the NH_2-group of substrate, amines and proteins. Hydrazones and azomethine derivatives of PALP have been used to help elucidate the activation mechanism. Since phospho-5'-pyridoximinotriazole (PPIT) was reported by GONNARD *et al.*, (1964) to be a potent activator of L-dopa decarboxylase (EC 4.1.1.26), it was compared with PALP in our present studies of a related enzyme, L-histidine decarboxylase (EC 4.1.1.22). While increase in PALP concentration resulted in an initial activation and subsequent inhibition, increase in PPIT concentration resulted in activation to a plateau with no inhibition at higher levels. At low concentration PALP was more active than PPIT in stimulating histidine decarboxylase activity, while at high concentrations the reverse was found. The activation pattern observed with PPIT was more consistent than that obtained with PALP, suggesting the suitability of this substance for the quantitative estimation of histidine decarboxylase in crude tissue preparations.

(Supported by MY-03663, NIH, and DAHC-19-67-G-0004, Army R&D)

GONNARD, P. *et al.* (1964). *Biochim. Biophys. Acta*, *81*, 548.

PHOSPHATASE ACTIVITY RELATED TO SYNAPTIC TRANSMITTER FUNCTION OF NORADRENALINE IN THE CENTRAL NERVOUS SYSTEM

Jørgen Clausen and B. Formby

The Neurochemical Institute, Copenhagen, Denmark.

A neutral Mg^{2+}-dependent p-nitrophenylphosphatase (p-NPP'ase) present in particulate fractions of rat brain was studied with regard to saturation with substrate, activation and/or inhibition caused by inorganic cations and catecholamines or derivatives of these.

Na^+ inhibited the p-NPP'ase activity, while K^+ specifically increased the Na^+ inhibited activity.

Noradrenaline and dopamine stimulated the p-NPP'ase. 3,4-Dimethoxyphenylethylamine decreased the noradrenaline stimulated p-NPP'ase. The effect of noradrenaline was influenced by K^+ and Na^+ in a way opposite to that seen in the K^+/Na^+ activated system without noradrenaline.

From the effects of 3-phosphoglycerate, α-glycerophosphate and ATP on the noradrenaline stimulated p-NPP'ase, it was possible to suggest the presence of two different enzyme systems (1) A K^+ activated and Na^+ inhibited p-NPP'ase (2) A Na^+ stimulated transphosphorylase or phosphotransferase.

On the basis of the demonstration of relative preponderance of phosphatidylserine in synaptosomes, the possible action of phosphatidylserine as a receptor of noradrenaline in the phosphatase system is discussed.

CHANGES IN BRAIN METABOLITES UNDER NORMAL PHYSIOLOGICAL CONDITIONS

J.A. Cocks and D. Richter

Medical Research Council Neuropsychiatric Research Unit, Carshalton, Surrey.

It is well established that the metabolism of the brain is influenced by treatments such as electrical stimulation or administration of convulsant drugs, but hitherto there have appeared only a few reports of changes in brain metabolites under normal physiological conditions. RICHTER & DAWSON (1948) found that the lactic acid content of the rat brain decreased during natural sleep and increased during emotional excitement. More recently GODIN & MANDEL (1965) have reported small changes in the free amino acids of the rat brain during sleep. The data now reported confirm the previous reports of changes in metabolites in the rat brain under physiological conditions and extend the previous observations on the rat to other animal species.

The concentration of lactic acid in the rat brain was found to vary with age: it increased from 0.88 μmoles/g at 1 day old to 3.14 μmoles/g at 60 days old. The adult hamster showed a similar level (3.11 μmoles/g) of brain lactic acid to that of the adult rat.

Lactic acid decreased from the adult value to about 1.97 μmoles/g in the hamster and in the rat brain during natural sleep. This decrease during sleep could be demonstrated in rats from 10 days to 60 days old. During pentobarbitone (Nembutal) anaesthesia for 1 hour in the rat and in the hamster, the brain lactic acid decreased to 1.96 and 1.13 μmoles/g, respectively. This decrease was not observed when the animals were kept in an incubator at 40° during the period of narcosis. The level of lactic acid in the rat brain increased to 8.74 μmoles/g in animals exposed to hypoxia for 20 minutes and allowed to recover.

The hamster was found to be a very suitable animal to use for sleep studies since it is less easily roused than the rat. Of the amino acids analysed, glutamic acid, alanine and GABA showed a small decrease in concentration in the hamster brain during undisturbed natural sleep of at least thirty minutes duration.

During hibernation in the hamster, the brain levels of threonine, alanine and GABA increased, whilst those of glutamic acid and lactic acid decreased.

GODIN, Y. & MANDEL, P. (1965). J. Neurochem., 12, 455.
RICHTER, D. & DAWSON, R.M.C. (1948). Amer. J. Physiol., 154, 73.

THE EFFECT OF X-IRRADIATION ON THE METABOLISM OF THE DEVELOPING CENTRAL NERVOUS SYSTEM

Stanley Cohan and Donald H. Ford

Department of Anatomy, State University of New York
Downstate Medical Center, Brooklyn, New York

We have undertaken a series of experiments to determine the effect of X-irradiation of neonatal rat brain on the metabolism of triiodothyronine (T_3) in the central nervous system (CNS) at different stages of postnatal development.

Male Wistar rats, receiving 100 roentgens X-irradiation solely to the head within twelve hours of birth, received 0.5 µg [^{131}I]T_3/kg body weight intravenously at one, two, three, or four weeks of age and were sacrificed $\frac{1}{2}$, 1, 2, 4 and 7 hours after injection. The percent of injected isotope present per gram of cerebrum, diencephalon, brain stem, cerebellum and plasma was compared to that in unirradiated rats at the same ages and time intervals.

In male Wistar rats similarly treated, but receiving 2.0 µg [^{131}I]T_3/kg body weight, CNS and plasma were extracted with ammoniacal ethanol for ^{131}I-containing compounds and chromatographed with n-butanol-dioxane-ammonia to determine the qualitative and quantitative distribution of ^{131}I.

Results demonstrated X-irradiation depression CNS T_3 uptake as measured by lower radioactivity levels in irradiated brain, and a difference in maturational changes in T_3-uptake between irradiated and control rats. Chromatographic analysis suggested X-irradiation depression of T_3 degradation rate, indicated by the higher percentage of total brain radioactivity associated with [^{131}I]T_3 in irradiated brains, and maturational differences in T_3 metabolism between irradiated and control rats. These changes reflect possible radiation-induced abnormalities in enzyme mediated systems responsible for thyroid hormone transport and degradation. Since many workers believe that the blood-brain barrier (BBB) may be in part a multi-faceted enzyme system responding to biochemical and physiological CNS needs, the system presently under consideration may be considered as part of the BBB. Thus, decreased levels of [^{131}I]T_3 in the X-irradiated CNS may be as indicative of damage to the BBB as increased CNS levels of metabolites frequently reported as resulting from BBB damage.

Supported by a grant from the Office of Naval Research, grant number NONR 4018 (01) (NR 101-592).

THE EFFECT OF CHANGES IN OXYGEN CONCENTRATION, ELECTROLYTE CONCENTRATION AND pH ON SUBSTRATE METABOLISM AND ULTRASTRUCTURE IN VIVO

Maynard M. Cohen and J.F. Hartmann

Department of Neurology, Presbyterian-St. Luke's Hospital, Chicago.

Previous studies indicated that oxygen lack, pH change and variation in electrolyte concentration of the medium produced profound changes in phosphorylation and ultrastructure of cerebral tissue incubated *in vitro*. Utilizing similar techniques, the effect of these conditions was studied on the oxidative decarboxylation of glucose and glutamic acid *in vitro* as well as on the ultra structural localization of sodium.

When tissue is incubated in the absence of oxygen and subsequently allowed to metabolize aerobically, the residual damage due to anoxia can be measured. Under these circumstances in addition to a decreased capacity to consume oxygen and maintain high energy phosphate concentrations there was a 25% diminution in the relative amount of CO_2 produced from uniformly labelled glucose. Aerobically, when the pH of the media was lowered to 6.4, oxidative decarboxylation was diminished a similar degree. When diminution in pH accompanied the anoxia, oxidative decarboxylation was subsequently decreased to less than 50% of normal with uniformly labelled glucose as substrate. Uniformly labelled glutamic acid differed from glucose under these circumstances in that the percent of CO_2 released from the incubated tissue remained unchanged both following anoxia alone and with accompanying acidosis.

A technique for precipitating tissue sodium *in situ* as a relatively insoluble and electron-dense compound (sodium antimonate) was applied to the electron microscopic study of the incubating brain slices. In these preparations, as well as in those from the cerebral cortex of normal intact animals, the sodium precipitate commonly appears as chains of fine particles, appearing in greatest numbers in astrocytes. In some micrographs, however, the precipitate takes the form of discrete circular masses of considerably larger diameter than that of the particles arranged in chains. The possible significance of these two sharply different forms of precipitate in relation to anoxia will be discussed.

FREE AMINO-ACID CHANGES IN CAT CEREBRAL CORTEX AFTER TOPICAL APPLICATION OF STRYCHNINE AND NEMBUTAL

Elisabeta Constantinescu

Neurological Institute of the R.S.R. Academy, Bucharest

Topical application of strychnine solution to the ectosylvian gyrus of the cat cerebral cortex, rapidly induces focal ECoG spikes; application of Nembutal induces focal ECoG hypersynchrony.

In chromatographic studies diffuse changes in some of the free amino-acids of the cerebral cortex were seen to occur even during the first 5 minutes after drug application. Such changes were seen in the ectosylvian gyrus and in the lateral and sigmoid gyri homo- and contralateral to the site of application. The changes were maximal after 5 minutes and returned to the control values at 30 minutes after drug application.

With both drugs, the γ-aminobutyric acid (GABA) level, showed practically no change for 30 minutes.

Strychnine application was accompanied by the following changes:

(1) a decrease in glutamic acid level, significant after 15 minutes

(2) a marked increase in aspartic acid, glycine, threonine and alanine after 5 minutes, reaching a maximum at 15 minutes after drug application

(3) a considerable increase in arginine level after 15 minutes

Topical application of Nembutal was accompanied by the following changes:

(1) a significant decrease in glutamic acid level which was maintained during 5 to 30 minutes of drug application

(2) a significant decrease in aspartic acid, glycine and threonine after 5 minutes

(3) a considerable decrease in arginine content after 15 minutes

(4) an increased content of lysine after 5 minutes

The curves giving amino acid concentration during 30 minutes of application of both drugs were analogous for all the six areas studied.

These and our previous data (CONSTANTINESCU & STERIADE, 1966) show again the importance of glutamic acid in connection with the excitability of the cerebral cortex.

RELATIONSHIPS BETWEEN INSOLUBLE PROTEINS IN MEMBRANE SUBCELLULAR FRACTIONS FROM RAT BRAIN

Carl W. Cotman, Walter J. Moore and Henry R. Mahler
Chemical Laboratory, Indiana University, Bloomington

The proteins of rat brain membranes viz., myelin, synaptic vesicles, and synaptosome membranes, were compared by means of disc electrophoresis and contrasted with those of mitochondria, and the microsomal and soluble fraction. The purity and composition of the fractions was established by electron microscope observation and by enzyme assays. NADH cytochrome c reductase, NADPH cytochrome c reductase, cytochrome oxidase, monoamine oxidase, and Na^+, K^+, Mg^{+2} stimulated ATPase were studied.

The protein of myelin separated into four bands, two of which had the solubility properties of proteolipids.

The protein composition of synaptic vesicles is complex, since they can be resolved into at least sixteen bands. Various microsomal subfractions, viz., microsomal membranes, microsomal vesicles, and microsomes with sedimentation properties similar to synaptic vesicles, display many similarities to synaptic vesicles. These fractions are distinct, however, on the basis of a few differences in band intensity or the presence of distinct bands. A single band was identified which had the solubility properties of a proteolipid.

Synaptosome membranes similarly separate into a complex spectrum of seventeen bands, each of which appear to be present in about equal concentration as approximated by staining intensities. The band pattern of synaptic vesicles, mitochondria, and microsomes resembles to some extent that of synaptosome membranes. The incorporation of label into individual protein bands from the subcellular fractions following an in vivo injection of amino acids will be presented and discussed.

PROTEIN LABELLING IN OCCIPITAL AND TEMPORAL CORTEX OF MONKEY

M. Cuénod, H.P. Metzger, A. Grynbaum and H. Waelsch

New York State Psychiatric Institute and
Columbia University, New York

The incorporation of tritium into proteins of various cortical fractions from 10 hemispheres of Macaca mulatta monkeys was studied in the course of an investigation on split brain preparation (life Sciences 5, 1115, 1966). Tritiated water (25 mc/kg) was injected intraperitoneally and, 4 hours later, the animals were decapitated. The occipital and temporal neocortex was dissected in each hemisphere and individually fractionated into synaptic endings, mitochondria, microsomes, nuclei, myelin and soluble fraction. The fractions were extracted with cold and warm TCA, ethanol, chloroform-methanol and ether, and the proteins, suspended in water, were freeze-dried repetitively in order to remove the freely exchangeable tritium; they were then combusted and the radioactivity was determined with a liquid scintillation counter.

The reliability of the method was determined by comparing the specific activities of corresponding fractions from symetrically treated homologous areas in a control animal: they were identical with a variability of $\pm 5\%$.

In each hemisphere, the specific activities of proteins in occipital and temporal cortex were different and showed a constant trend: the synaptic endings were $19 \pm 8\%$ and the microsomes $23 \pm 7\%$ higher in the occipital cortex; the nuclei were $12 \pm 9\%$ lower in the occipital cortex than in the temporal, and the mitochondrial, myelin and soluble fractions were the same in both lobes.

These results suggest that protein sunthesis, or the migration of newly formed proteins, have a regional characteristic which might correspond to a functional spcificity.

THE LIPIDS OF HUMAN PINEAL GLAND

J. Czarnocki, P.S. Sastry and Harvey C. Stancer

Neurochemical Section and Department of Psychiatry, Clarke Institute, University of Toronto

The pineal gland has been known to man since the time of Galen and has been studied by numerous investigators. Still our knowledge of this organ is incomplete and its function is obscure. Recent reports suggest that the pineal body is part of the limbic system and through its neurotransmitter melatonin, exerts an important effect on sexual maturation (RELKIN, 1966; WURTMAN & AXELROD, 1965). Although considerable morphological information is available (KAPPERS & SCHADE, 1965) relatively little is known about the chemical composition of the pineal gland. PROP (1965) and JOUAN *et al*. (1964) have reported on some aspects of the lipid composition of rat and hog pineals respectively. We have chosen to extend this work by determining the complete lipid composition of pineal glands obtained from different mammalian species, with particular emphasis on the human.

The glands from humans, who had died suddenly, were obtained from 3 to 22 hours after the event. These were compared with beef and sheep pineals obtained within one hour of death. Pooled samples of each species were extracted by the procedure of FOLCH *et al*. (1951). The extract yielded 9.2, 5.8 and 10.8% total lipid of wet weight of tissue; total phospholipid phosphorus of 665,896 and 960 micrograms per gram wet weight from human, beef and sheep glands, respectively. Cholesterol was found only in its free form in concentrations of 3.64, 6.09 and 5.65 mgm per gram fresh tissue for human, beef and sheep respectively. Silicic acid impregnated paper chromatography indicated that the phospholipid composition in all three species is comparable. The main constituents were lecithin, sphingomyelin, phosphatidyl inositol and phosphatidyl serine whereas phosphatidyl ethanolamine and cardiolipin were present as minor constituents. In glands obtained from humans of varying age the greatest variation in phospholipid composition was found in the phosphatidyl ethanolamine fraction. The variations and the results of fatty acid analysis will be discussed.

FOLCH, J., ASCOLI, I., LEES, M., MEATH, J.A., LEBARON, F.N. (1951). *J. biol. Chem.*, 191, 833. JOUAN, P., VIEM DAI, T. & CORMIER, M. (1964). *Bull. Soc. Chim. Biol.*, 46, 1121. KAPPERS & SCHADE. (1965). *Structure and Function of the Epiphysis Cerebri*. Elsevier, Amsterdam. PROP, N. (1965). In *Structure and Function of the Epiphysis Cerebri*. Edited by Kappers & Schade. Elsevier, Amsterdam. RELKIN, R. (1966). *New Eng. J. Med.*, 274, 944. WURTMAN, R.J. & AXELROD, J. (1965). *Sc. Amer.*, 213, 50.

THE INTRANEURONAL DISTRIBUTION, TRANSPORT AND LIFE-SPAN OF NORADRENALINE STORAGE GRANULES IN THE ADRENERGIC NEURON SYSTEM

Annica Dahlström

Department of Histology,
Karolinska Institutet, Stockholm

The intraneuronal distribution of noradrenaline (NA) in the peripheral adrenergic neuron system of some mammals (rat, cat, rabbit) was investigated. It was found that the ganglion cell body contains on the average 0.4 pg of NA, and that the nerve terminal system of each neuron contains about 300 times the content of the cell bodies, i.e. 0.12 ng of NA (DAHLSTRÖM & HÄGGENDAL, 1966). The summarized length of the terminal arborizations of one neuron was calculated to be about 10-30 cm in the rat and the cat. Each varicosity of the adrenergic nerve terminals was found to contain 5×10^{-3} pg of NA on the mean (DAHLSTRÖM, HÄGGENDAL & HÖKFELT, 1966).

Several facts support the idea that the NA in adrenergic nerves of normal mammals is stored for the major part in the specialized storage granules (for discussion see i.a. MALMFORS, 1965; DAHLSTRÖM, 1966). Therefore it seems reasonable to assume that the amount of NA in the different parts of the neuron corresponds to a certain number of storage granules. It was suggested on the basis of certain assumptions (see DAHLSTRÖM, HÄGGENDAL & HÖKFELT, 1966) that an average varicosity of the adrenergic nerve terminals would contain about 1500 amine storage granules.

The storage granules are synthesized in the perikarya of the neurons and transported down the axons to the terminals. If the axon is constricted NA storing granules are accumulated proximal to the lesion, the amount of accumulated granules increasing linearly with the time after ligation. The rate of the proximo-distal transport of granules was calculated to be about 5.10 and 3 mm per hr in the rat, cat and rabbit sciatic nerve, respectively. The life-span of granules in the terminal system was calculated to be about 35, 70 and 50 days in the above mentioned species (DAHLSTRÖM & HÄGGENDAL, 1966, 1967).

The results do not indicate a release of whole granules on nerve stimulation, since it was found that the turn-over time of NA in the nerve terminals is about 1-2 days (BJÖRLING & WALDECK, 1965).

BJÖRLING, M. & WALDECK, B. (1965). _Life Sci._, _4_, 2239. DAHLSTRÖM, A. (1966). M.D. Thesis, Stockholm. DAHLSTRÖM, A. & HÄGGENDAL, J. (1966). _Acta physiol. scand._, _67_, 271. DAHLSTRÖM, A. & HÄGGENDAL, J. (1966). _Acta physiol. scand_. _67_, 278. DAHLSTRÖM, A. & HÄGGENDAL, J. (1967). _Acta physiol. scand_. (in press). DAHLSTRÖM, A., HÄGGENDAL, J. & HÖKFELT, T. (1966). _Acta physiol. scand._, _67_, 287. MALMFORS, T. (1965). _Acta physiol. scand._, _64_, Suppl. 248.

EXOPEPTIDASE ACTIVITIES OF BRAIN SUBCELLULAR MEMBRANE FRACTIONS

R.K. Datta, N. Marks and A. Lajtha

New York State Research Institute for Neurochemistry and Drug Addiction, Ward's Island, and College of Physicians and Surgeons, Columbia University, New York, N.Y.

The study of some of the properties of a number of brain proteinases and peptidases recently separated in our laboratory enabled us to follow the distribution of the enzymes in various subcellular fractions of brain. The involvement of some membranes in the cell in protein synthesis and the formation of α-chymotrypsinogen by membrane-attached ribosomes of pancreas (SIEKEVITZ & PALADE, 1960) emphasize the importance of studying those enzymes in ribosomes and in microsomal membranes that may have a role in protein and peptide breakdown. For this reason, peptidases were studied in cerebral microsomal fractions obtained by the treatment of these fractions with Triton X-100 or GR-5, desoxycholate, and digitonin. As previously reported (MARKS & LAJTHA, 1963), microsomal membrane fractions and ribosomes showed no neutral or acid proteinase activity when tested with haemoglobin or casein as substrate. However, membrane fractions yielded high aminopeptidase activity with marked specificity towards L-leu-gly-, L-leu-gly-gly, and L-ala-gly-gly, and showed also significant arylamidase activity in the present of arg-, lys-, meth-, and leu-β-naphthylamides. In contrast, ribosomes showed only trace aminopeptidase and arylamidase activities with these substrates, which indicated that most of the enzymic activity in the microsomal fraction resided in the membrane portion. On the basis of substrate specificity and enzyme stability, effects of SH- and oxidizing agents. and pH optima, indications were obtained of the presence of a multiple enzyme system for peptide breakdown in the microsomal membrane fraction. Membrane-bound aminopeptidases and arylamidases could be purified and the properties studies, following their extraction with 0.5% desoxycholate, dialysis, and repeated passage through DEAE and CM columns. Membrane arylamidases are activated by Co^{2+}, Mn^{2+}, and Ca^{2+}, inhibited by puromycin (Ki 8 x 10^{-4} M), and have a pH optimum of 6.5 and a Km of 7 x 10^{-4} M. Amino peptidases (tested with L-leu-gly-gly) are not activated by these metal ions and not inhibited by puromycin. The pH optimum of membrane aminopeptidase (leu-gly-gly) was 7, with a Km of 6 x 10^{-4} M. The presence of peptidases in those membrane fractions which are known to be involved in protein synthesis suggest that these membranes are also involved in protein turnover as well as in transport processes.

MARKS, N. & LAJTHA, A. (1963). Biochem. J., 89, 438.
SIEKEVITZ, & PALADE, (1960). J. Biophys. biochem. Cytol., 7, 619.

THE SUBUNIT STRUCTURE OF NEUROFILAMENT PROTEIN

P.F. Davison, F. Huneeus-Cox, D. Lusted and F.O. Schmitt

Massachusetts Institute of Technology, Boston.

Neurofilaments were isolated from the axoplasm of giant axons of the squid Dosidicus Gigas by zone sedimentation through density gradients. The protein is very sensitive to oxidation and must be protected by the addition of thiol compounds. The subunit proteins comprising the neurofilaments may also be selectively aggregated by freezing. Neurofilament protein obtained by either of these procedures could be dissolved in urea or thiocyanate and alkylated by iodoacetate or ethyleneimine. Preparations so treated were water soluble but contained aggregated material. Succinylation effected a further dissociation of the protein subunits; these preparations were studied by ultracentrifugation, gel electrophoresis, and by immunological methods. The relationship between protein subunits, their ready aggregation and the structure of the neurofilament as well as the immunological cross-reactivity of neurofilaments from species other than squid will be discussed.

ON A NEW GROUP OF ETHANOLAMINE CONTAINING PHOSPHOLIPIDS

H. Debuch and G. Wendt

Physiologisch-chemischen Institut der Universität,
Köln

Both from the FOLCH-extract and from the acetone-extract of fresh ox brain we were able to isolate a hitherto unknown fraction of glycerophospholipids. This fraction consists of nearly 60% of the plasmalogen form. Mild acid treatment produced the lysoderivative and the diesterform of the original substance, which showed a R_F value higher than that of phosphatidyl-ethanolamine. By treatment under mild alkaline conditions the fatty acids were split off and the resulting water-soluble phosphate esters were glycerylphosphoryl-ethanolamine and an unknown one, which appears to be the lyso-ethanolamine plasmalogen.

The latter compound has also been detected in the organic phase after mild alkaline hydrolysis. This shows that the new group of phospholipids consists of derivatives of phosphatidylethanolamine.

The ethanolamine derivative was isolated and investigated: the results will be discussed.

GANGLIOSIDE ANALYSES OF SERIAL SECTIONS THROUGH BRAIN REGIONS AND OF ISOLATED NEURONES

D.M. Derry and L.S. Wolfe

Donner Laboratory of Experimental Neurochemistry, Montreal Neurological Institute, McGill University, Montreal

There is good evidence that brain gangliosides are entirely localized in neuronal membrane structures and may be used to indicate the abundance of neuronal membranes in various cortical laminae. The ganglioside content of 24μ thick serial cryostat sections through area CA2 of the ox hippocampus cut parallel to the ventricular surface and through cerebellar folia cut parallel to the pial surfaces was determined by a fluorimetric method. Wide differences in ganglioside content even of adjacent sections was found within the various laminae, particularly in the molecular and granular layers. In the hippocampus the lowest values (0.82 mg/g dry wt.) occurred in the alveus and the highest in the granular and polimorph layers of the fascia dentata (18.2 mg/g. dry wt.). In the cerebellum the ganglioside contents of individual sections in the molecular and granular layers varied from 4 to 15 mg/g dry wt. A fall in ganglioside content occurred at the level of Purkinje cell bodies. The patterns of ganglioside distribution in tissue blocks from the same regions in different animals were not superimposable. The wide differences in ganglioside content found in individual tissue blocks within the granular and molecular layers of the hippocampus is not artefactual since these variations disappear when serial sections were cut so as to include all the structural areas in each section. The results indicate that gangliosides are not uniformly distributed throughout the neuronal membrane. The differences in ganglioside content within the various laminae may be related to differences in the concentrations of dendritic membranes and synapses. Ganglioside contents were also determined in groups of neurones and glial tissue dissected from Deiter's nucleus of the brain stem of the ox. The mean ganglioside content of clean isolated neurones was 11.8 mg/g dry wt. However, the neuropile material trimmed immediately surrounding the neurones was also rich in gangliosides. The ganglioside content of isolated neurones together with the untrimmed glial tissue was 4.6 mg/g dry wt. These results indicate that gangliosides may also occur in nerve terminals.

GLUTAMINE SYNTHETASE IN THE BRAIN

Christa Dierks-Ventling

Rosewood State Hospital, Owings Mills, Maryland 21117

Glutamine synthetase (GS) of rat liver was twice as active as that of rat brain when 0.25 M-sucrose homogenates were assayed for 10 min. at 37°C under conditions standardized by us. Repeated freezing and thawing increased the activity of the brain GS, but not of the liver GS. When the homogenates were prepared with 0.25 M-sucrose containing 0.6% desoxycholate (DOCH), and dialysed prior to the assay, the brain enzyme increased three times while the liver enzyme did not change significantly. DOCH was found to be more efficient than tween or sodium lauryl sulphate at the same concentration. After 20 min. incubation at 37°C the brain enzyme lost activity. This could be counteracted by the addition of 0.04 M-citrate. The effect was found for both brain and liver, whether DOCH was added to sucrose-homogenate or not, but the effect was most conspicuous with the brain DOCH homogenate. Km studies showed that the value obtained by MEISTER (1962) for crystalline sheep brain enzyme or for partially purified rat liver enzyme by WU (1965) could be reproduced with whole homogenates of brain or liver. DOCH did not alter the Km, but increased Vmax considerably, suggesting a release of otherwise bound enzyme. Brain fractionation showed that brain mitochondria do contain GS and that the extra activity obtained through DOCH action is above that obtained from microsomal contamination. Liver mitochondria do not contain GS. Homogenates of developing rat brain show an increased activity with DOCH before the fourth day after birth and this increase continues and becomes more pronounced daily until weaning time when it reaches adult level. The developing liver by contrast shows no difference. The sensitivity of brain to ammonia poisoning may be related to its ability to convert toxic ammonia to innocuous glutamine. These findings may be a basis for study of the relative toxicity of ammonia in the newborn and adult.

MEISTER, A. (1962). The Enzymes, 6, 443. WU, C. (1965). Biochim. Biophys. Acta, 96, 134.

THE EFFECTS OF UNDERNUTRITION ON MYELINATION AND ITS LATER SEQUELAE

J. Dobbing

Institute of Child Health, London, W.C.1.

Some of the effects of undernutrition on the brains of developing rats and pigs have already been reported (DOBBING, 1964; 1966). Observations have now been extended for longer periods in rats, and to include more detailed lipid analysis of the cerebrum and spinal cord in pigs.

Groups of rats were underfed at various times during development, including periods from birth to three weeks, birth to eleven weeks and from three to eleven weeks. They then were allowed food *ad libitum* until twenty-eight weeks of age, when they were killed. DNA-P and cholesterol were used as indices of cell numbers and myelin respectively. Results suggest that small but demonstrable deficits are present in the mature brain as a result of comparatively mild undernutrition confined to the first three weeks of life, which is the period of fastest growth. Very severe undernutrition of *adult* rats, reducing the body weight to almost half its former value in five weeks, has had no effect whatever on either the brain weight or any of the brain lipids separated by thin-layer chromatography.

In the forebrain of normally developing pigs, lipids have been classified into three main groups: (a) cerebroside and cholesterol, whose concentration per gram wet weight has not reached its adult value by one year of age, and which continues to increase for at least 3 years: (b) total phospholipids, serine and inositol phospholipids, ethanolamine phospholipids and sphingomyelin, whose concentration in fresh tissue has reached an adult value by one year of age: (c) choline phospholipids whose adult concentration is achieved by 4 weeks of age. When pigs were severely undernourished from two weeks of age to one year, and then rehabilitated on an *ad libitum* diet until they were three and a half years old, the lipids of groups (b) and (c) were found to have completely recovered, equivalent to concentration values found in normal pigs of the same age, although the absolute size of the brain was reduced. The concentrations of cerebroside and cholesterol, however, remained deficient. The possibility arises that myelination may be a process specifically vulnerable to undernutrition at the time of its fastest rate.

DOBBING, J. (1964). *Proc. Roy. Soc. B.*, 159, 503.
DOBBING, J. (1966). *Biol. Neonat.*, 9, 132.

ACETYLCHOLINE AND SOME FACTORS AFFECTING THE PERMEABILITY OF CELL MEMBRANES IN NERVOUS TISSUES

Nikolai N. Doemin

Laboratory of Functional Neurochemistry
I.P. Pavlov Institute of Physiology, Academy of Sciences of the USSR, Leningrad.

Acetylcholine (ACh), apart from acting as a transmitter in the transfer of activity in cholinergic nervous structures, also affects a number of intracellular biochemical processes in nervous and in other tissues. ACh seems to be included in the system of fine control of cell metabolism.

The present study deals with the influence of ACh on certain factors known to affect the permeability of cell membranes: these include the Na^+-K^+-activated ATP-ase in different subcellular fractions of rat brain (experiments by Dr. Yevgueniya N. Makarova) and the metabolism of γ-aminobutyric acid (GABA) in brain homogenates (experiments by Dr. Nataliya S. Nilova).

It has been shown that ACh at concentrations from 0.6 x 10^{-6} to 0.6 x 10^{-2} M does not influence the Na^+-K^+-activated ATP-ase in the nuclear and purified mitochondrial fractions, but inhibits its activity in the microsomal fraction: 0.6 x 10^{-5} M-ACh by 5%, 0.6 x 10^{-4} M-ACh by 12%, 0.6 x 10^{-3} M-ACh by 20%, and 0.6 x 10^{-2} M-ACh by 30%. A slight inhibition of the Na^+-K^+-activated ATP-ase also appears to be produced by 0.6 x 10^{-3} M- and 0.6 x 10^{-2} M-ACh in preparations of membrane fragments sedimenting after differentional centrifugation of the nuclear fraction (by 9 and 13% respectively) and of the mitochondrial fraction (by 8 and 20% respectively).

ACh at a concentration o.6 x 10^{-4} M inhibited by about 25% the activity of the GABA-aminotransferase: it had no effect on glutamate decarboxylase activity. Thus, ACh was able to reduce the rate of elimination of GABA — a factor which could increase the permeability of the cell membranes.

ALTERATION OF PHOSPHOLIPIDS IN WALLERIAN DEGENERATION

J. Domonkos and L. Heiner

Brain Research Institute, University Medical School, Szeged, Hungary

The biochemical processes underlying myelin destruction can be most favourably detected by studying the chemical changes occurring in Wallerian degeneration. In the present study the phospholipids of the sciatic nerve of the rabbit were submitted to chemical analysis for a period of 14 days in the early stage of Wallerian degeneration. It could be established that lipid-P and fatty acid esters underwent a successive parallel decrease, attaining a reduction of about 40-50 per cent by the 14th day. The glycerol phosphatides containing fatty aldehydes (plasmalogens) showed a reduction that was parallel to that of lipid-P until the 6th to 8th day: at the 14th day of observation about 60-70 per cent of plasmalogens had been destroyed. The ratio of fatty acid ester to lipid-P thus remained unaltered in the first two weeks of Wallerian degeneration and the ratio of fatty aldehyde to lipid-P diminished markedly from about the 6th to 8th day onwards.

The values of the ratios suggest that in the diacyl glycerolphosphatides the two ester linkages are generally destroyed to the same extent: therefore from these phosphatides there is no possibility of accumulating lysophosphatides. However the vinyl ether linkages are destroyed more quickly than the ester linkages so that the decomposition of the plasmalogen molecule will only be partial: therefore lysophosphatides can be accumulated through the decomposition of phosphatides containing fatty aldehydes. It could not be established that there is an accumulation of lysophosphatidylcholine by the 14th day after the nerve section. On the other hand, the amount of lysophosphatidylethanolamine (distribution per cent in phospholipid fraction) increased at the same time. Fatty aldehydes occur in ethanolamine phosphatides while the other phosphatides contain fatty aldehydes only in traces in the sciatic nerve: therefore the decomposition of the phosphatidal ethanolamine is due to a relative accumulation of lysophosphatidylethanolamine.

ELABORATION AND TRANSPORT OF AXONAL PROTEINS AS REVEALED BY ELECTRON MICROSCOPE RADIOAUTOGRAPHY

B. Droz

Département de Biologie, Centre d'Etudes Nucléaires, Saclay

The axon is an elongated nerve cell process which does not possess ribosomes. Previous radioautographic data have shown that the axon is not able to synthesize its own proteins and receives continuously new proteins manufactured by its cell body. This cellulifugal migration characterizes the turnover of the axonal proteins (DROZ & LEBLOND, 1963).

Electron microscope radioautography was used to identify the cell organelles in which the axonal proteins are elaborated and transported. The concentration of radioactivity was measured in the organelles of spinal ganglion cells at various times after intravenous injection of [^{3}H]leucine or [^{3}H]lysine in rats. The earliest peak of radioactivity appeared in the ribosome-rich Nissl substance (2-5 min); other peaks were observed later in the smooth-surfaced membranes of the Golgi complex (20-30 min), in the neuroplasm (10 min - 2 hr) and in the mitochondria (10-30 min). Next, labelled organelles were seen to migrate cellulifugally from the axon hillock (30 min) into the proximal regions of the myelinated axon (2 hr). The radioactivity found in the axons was mostly related to filaments and tubules present in the axoplasm. The other components of the axon (mitochondria, vesicles and axolemmal membrane) accounted for only a minor part of the labelling. Our results indicate that:

(1) most of the axonal proteins by-pass the Golgi region and may contribute to the formation of the axoplasmic neurofilaments and neurotubules;

(2) proteins which accumulate in the Golgi complex give rise to migratory vesicles;

(3) the nerve cell body dispatches into the axon mitochondria containing newly-formed proteins.

In conclusion, the various proteins, which share a common pathway of axonal migration, appear to originate from and migrate into a series of different cell organelles.

PAPER CHROMATOGRAPHIC SEPARATION AND QUANTITATIVE ANALYSIS OF CERTAIN B_6 COMPONENTS AS THEIR DANSYL DERIVATIVES

I. Durkó* and A.A. Boulton

Brain Research Institute, University of Szeged, Hungary, and Medical Research Council Unit for Research on the Chemical Pathology of Mental Disorders, The Medical School, Birmingham, 15.

Many workers have postulated that a disturbance in the function of the pyridoxal enzymes occurs in certain forms of mental illness (schizophrenia, phenylketonuria, epilepsy, etc.). HUSZÁK and DURKÓ (1964) have shown that in the active phase of schizophrenia an increase in the excretion of pyridoxal and of certain indole metabolites, some in the form of an indole-pyridoxal complex, occurs.

In order to measure the amounts of pyridoxal and pyridoxamine, as usually extracted from blood, directly on paper chromatograms, these substances were first converted to fluorescent derivatives by reaction with the Dansyl reagent (1-dimethylaminonaphthalene-5-sulphonyl chloride) as described by GRAY and HARTLEY (1963) and BOULTON and BUSH (1964). The most effective separation of these derivatives was achieved by chromatography overnight on Whatman No. 4 paper in the solvent system light petroleum:toluene:acetic acid:water, 133:66:170 30 (v/v). After drying at room temperature, direct evaluation of the chromatograms was obtained using a paper strip scanning device arranged for fluorescent measurements (activation 3650 Å, Chance coloured glass filter OX1, fluorescence 5000 Å Chance OY8). The derivatives are not appreciably light sensitve (fluorescence loss less than 10% over 3 days) and produce linear calibration curves in the range 0 - 5 μgm. By heating the chromatographic strips prior to scanning (BOULTON, 1965) a substantial increase in sensitivity ensues so that sub-microgram quantities can be measured.

Because of its extremely low concentration in blood the separation and analysis of pyridoxin was not attempted.

*I am grateful to the Wellcome Foundation for the provision of a Research Travel Grant.

BOULTON, A.A. (1965). Federation of European Biochemical Societies, 2nd Meeting, Vienna, A.265. BOULTON, A.A. & BUSH, I.E. (1964). Biochem. J., 92, 11P. GRAY, W.R. & HARTLEY, B.S. (1963). Biochem. J., 89, 59P. HUSZÁK, I. & DURKÓ, I. (1964). Acta Biochem. Polon., 11, 389.

AMPHETAMINE AND FREE NUCLEOTIDES IN THE BRAIN

S. Edel

Centre de Neurochimie du C.N.R.S., Strasbourg

Amphetamine, injected in mice, induces motor hyperactivity, especially evident when the animals are confined to a small space (CHANCE, 1946, 1947).

It is highly probable that motor hyperactivity is associated with an augmentation of brain metabolism and an increased utilization of energy rich phosphates. We have observed that the brain ATP level remains unmodified or, in some cases, even increases, after injection of 15 mg of amphetamine per kg body weight. The ATP/ADP ratio also remains unchanged. It appears that an increased synthesis of energy rich phosphates adequately compensates for the increased energy demand during hyperactivity. In order to verify this hypothesis, we studied the incorporation of ^{32}P into mouse brain phosphocreatine and ATP-α, -β and -γ phosphates. Control and amphetamine-treated animals were sacrificed at the same time after injection of $H_3{}^{32}PO_4$. One group of amphetamine-treated animals was sacrificed at 1 hr, another group 3 hr, a third group 6 hr after administration of the drug. The specific activity of the organic phosphates was expressed as relative specific activity (RSA), that is: specific activity relative to that of the inorganic phosphate in the same brain.

The specific activity of brain inorganic phosphate of drug treated animals was always lower, while the RSA of brain phosphocreatine was always higher than in untreated animals. In the animals sacrificed 1 hr, 3 hr and 6 hr after amphetamine injection, the RSA of brain ATP-γ phosphate was respectively 11, 21 and 66 per cent higher than in untreated mice. The corresponding values for the β-phosphate were 2, 10 and 47 per cent and for the α-phosphate 19, 35 and 55 per cent more.

Our results indicate that oxidative phosphorylation and energy production adequately increase during central nervous system hyperactivity.

CHANCE, M.R.A. (1946). J. Pharmacol., 87, 214. and (1947) J. Pharmacol., 89, 289.

RESPIRATORY CONTROL OF NORMAL AND OF β-INTERNAL AND γ-EXTERNAL IRRADIATED CNS MITOCHONDRIA

Enrique Egana

Laboratory of Neurochemistry, Institute of Experimental Medicine, School of Medicine, University of Chile, Santiago, Chile

It has been shown that low doses of β-internal irradiation enhance the '*in vitro*' respiration of CNS slices taken from various areas of the neuraxis (from brain cortex to spinal cord).

The aim of this investigation was to study the influence of β- and γ-exposure on respiration and on the activity of the components of ATP synthesis, i.e. the phosphate acceptor ADP and P_i, of CNS mitochondria. Experiments were devised to study the best procedure to obtain the mitochondrial preparations (ALDRIDGE and WHITTAKER methods). When these had been obtained from brain cortex, diencephalon, hypothalamus, olfactory bulbs, etc., the irradiated mitochondria were compared with normal. Various '*in vitro*' respiration media were used, the basic substances being: succinate, ATP or ADP and β-NADH. The effect of glucose, hexokinase, cytochrome-b, glutamate and Ca^{++} were also studied. The results are expressed μmoles O_2 or μmoles substrate utilized per mg mitochondrial protein.

Determinations of the substrates (succinate, glucose) used and the ATP formed or split off and the ADP and P_i used, completed this study. The general results showed that respiration was inhibited by high doses of exposure. The effect depended on the type of radiation. The synthesis of ATP was different in normal and irradiated nervous tissue.

CHANGES IN THE PROTEINS OF HUMAN AND RABBIT NEURAL TISSUE WITH DEVELOPMENT

Elizabeth Roboz Einstein and Judit Csejtey

Department of Neurology, University of California School of Medicine, San Francisco.

A few studies have appeared on protein composition of the brain correlated with developmental age, using solvent systems which would not solubilize the basic proteins present in ribosomal, nuclei, and myelin fractions.

In the present studies, human and rabbit brain and cord from embryo to full maturity were used. Since myelin deposition is an important phase of development, it was included in these studies. The myelin was prepared according to NORTON (personal communication, 1966). For extraction of the basic (cathodal) proteins, HCl with pH 1.2 - 2.0 with simultaneous or prior removal of the lipids by organic solvents was used; while for the anodal proteins, glycine buffer of pH 9.6. For complete solubilization of neural tissue, phenol-acetic acid water was used without apparent denaturation (SYNGE and coworkers used this solvent for leaf proteins).

To study the proteins in developing brain, acrylamide electrophoresis, which has the greatest resolving power, was employed. Rigorous studies were carried out in both disc and flat-form, using different concentrations of the gel with lanthanum acetate, or urea, or phenol-acetic acid. For the separation of anodal proteins, the disc and flat (slab) type were equally satisfactory; but for the separation of basic proteins, the flat type proved superior.

The number of basic proteins in the white matter is much greater than in the myelin because of the presence of other basic proteins. Certain proteins with the same electrophoretic mobility as the encephalitogen (see NAKAO et al., 1966) are not present in the early stages of development. With the appearance of this particular protein, other basic proteins greatly diminish. Since the appearance of the encephalitogen is gradual and well discernible by electrophoresis, it is proposed that this protein be used as a yardstick for measuring development.

Supported by USPHS and National Multiple Sclerosis Society grants.

NAKAO, et al. (1966). Biochim. Biophys. Acta, 130, 163 & 179.

THE MATURATION OF HUMAN CNS MYELIN

L.F. Eng, B. Gerstl, D.V. Pratt and M.G. Tavaststjerna

Department of Pathology, Stanford University, Stanford, Cal. and Research Laboratory, Veterans Administration Hospital, Palo Alto, California.

Myelin was prepared from human white matter from individuals of age 10 weeks, 4 years and adult. The latter included a case of multiple sclerosis (MS). By selective solvent extraction and ultracentrifugation three major fractions of myelin were obtained. A mixture of proteolipid protein (A) and basic protein (B) was extracted from myelin with a solution containing 0.5 M ammonium acetate and 0.5% Triton X-100. The proteolipid protein (A) was precipitated by partition with diethyl ether. While total protein and lipid content of myelin at the different ages were similar, the distribution of the various proteins and lipids changed with age.

The basic protein (B) increased from 16% of the total myelin protein at 10 weeks to about 30% in the adult. Amino acid composition, molecular weight (9000), and experimental allergic encephalomyelitic activity of the basic protein were similar for normal adult and MS myelin. The proteolipid protein (A) decreased from 68% of the total myelin protein at 10 weeks to about 50% in the adult. The amino acid composition of an adult sample was similar to that of the classical proteolipid protein of FOLCH. The third fraction (C) which contained the bulk of the myelin lipids contained 17 to 20% of the myelin protein at the various ages.

The composition of the total myelin lipids changed with age as reported by GERSTL _et al_. (1966) and as also found for the rat (L.F. ENG, E. NOBLE and T. KOK, unpublished data). Among the phospholipids, lecithin decreased, sphingomyelin increased with increasing age, while the serine and ethanolamine phosphatides remained relatively constant. The proportions of galactolipids and cholesterol to phospholipid increased with age. The phospholipids associated with the proteolipid protein (A) did not change with age. The development of the lipids associated with the third protein fraction (C), however, paralleled that of the total myelin lipids.

GERSTL, B. _et al_. (1966). _Arch. Neurol._, _15_, 603.

DISTRIBUTION AND EXCHANGE OF BRAIN CALCIUM *IN VIVO* IN THE RAT

Farouk F. Faragalla

Cornell University Medical College, New York

The distribution of calcium was studied in the following brain areas: pituitary gland, hypothalamus, pons, medulla oblongata, brain stem, cerebellum and grey and white matter or cerebrum. Calcium content was determined in tissue extracts according to AMES & NESBETT (1958). A wide range of calcium concentrations was found, with the pituitary gland containing 74 $\pm$ 10 meq of Ca/kg wet tissue and the cerebral cortex containing 4 $\pm$ 0.5 meq/kg. The calcium content of the rest of the tissues fell in between.

The distribution of extracellular and intracellular calcium was derived according to LOWRY & HASTINGS (1942). There was no difference in the extracellular calcium distribution in any tissue except in the pituitary gland where extracellular calcium was much less. The distribution of intracellular calcium paralleled distribution of total calcium in the various brain areas. The possible significance of this heterogenous distribution of calcium regarding function will be discussed.

Exchangeability of brain calcium has been determined after intraperitoneally administering a tracer dose of ^{47}Ca to rats. Injected animals were sacrificed after 10 minutes, 1, 3, 6, 12, 18 and 24 hours. Uptake of radioactive calcium by brain tissues increased gradually reaching its maximum at 3 hours, except in the case of the pituitary where maximum radioactivity was reached after 1 hour.

Biological half life of the radioactive calcium varied between 1.5 hours in the pituitary gland to 20 hours in the cerebral cortex. The biological half-life of radioactive calcium in other tissues fell in between.

Exchangeable calcium was calculated by dividing the radioactivity appearing in the tissue by the integrated serum specific activity curve. The exchangeability of calcium varied; ranging between 10-30% of the total calcium content. Equilibrium between brain and serum was reached in 6-12 hours after injection.

AMES, A & NESBETT, F.B. (1958). *J. Neurochem.*, *3*, 116.
LOWRY, O.H. & HASTINGS, A.B. (1942). *J. biol. Chem.*, *143*, 257.

THE SYNTHESIS OF PHOSPHOENOLPYRUVATE BY MITOCHONDRIA OF MAMMALIAN BRAIN CORTEX

R.A. Felicioli and C.A. Rossi

"Impresa Enzimologia" of the C.N.R.,
Institute of Biological Chemistry, University of Pisa,
Pisa, Italy

Evidence for the carboxylation of pyruvate in the mammalian brain was obtained by measuring the incorporation of $^{14}CO_2$ administered to cats by intracarotid infusion. Soon after the infusion the highest specific activity in cerebral extracts was found in aspartic acid. Products obtained from ^{14}C-labelled pyruvate also indicated the formation of oxaloacetate by CO_2 fixation. After administering [2-^{14}C]pyruvate to rats intracerebrally, isolation of ^{14}C-products showed much accumulation of ^{14}C in the free amino acids. Labelling in 2-C or 3-C of glutamate implies the utilization of pyruvate through the carboxylation pathway.

The carboxylation of pyruvate is catalysed by the biotin-containing enzyme pyruvate carboxylase (E.C.6.4.1.1.) which occurs in the mitochondria of many animal tissues. In experiments with preparations of brain cortex we observed the presence of pyruvate carboxylase in the mitochondria of different mammalian species. The finding of the pyruvate carboxylase in the brain cortex in appreciable amount raises the question of the role of the CO_2-fixing enzymes in the C.N.S. It may be suggested that CO_2 fixation contributes in this tissue to the synthesis of the dicarboxylic acids needed for the operation of the Krebs cycle. The presence of the pyruvate carboxylase in the mammalian brain cortex can explain the findings of WAELSCH *et al*. (1964) and of MCMILLAN and MORTENSEN (1963) on the contribution given to the tricarboxylic acid cycle by CO_2 fixation in the brain cortex.

Evidence has been presented suggesting that phosphoenolpyruvate is synthesized from pyruvate or lactate in the overall process of gluconeogenesis through a pathway including the pyruvate carboxylase and phosphoenolpyruvatecarboxykinase (E.C.4.1.1.32) functioning in sequence. We also found the latter enzyme in the brain cortex mitochondria of different mammalian species.

Our data demonstrate the potentiality of mammalian brain cortex mitochondria for synthesizing phosphoenolpyruvate from pyruvate or lactate.

MCMILLAN & MORTENSEN (1963). *J. biol. Chem.*, *238*, 91.
WAELSCH *et al*. (1964). *J. Neurochem.*, *11*, 717.

A SENSITIVE CHEMICAL METHOD FOR THE DETERMINATION OF ACETYLCHOLINE; ITS APPLICATION IN THE STUDY OF PRESYNAPTIC RELEASE

J.H. Fellman

University of Oregon Medical School, Portland, Oregon.

Our earlier work which demonstrated an active acyl character to acetylcholine (ACh) led us to hypothesize that ACh acts biologically by an initial acyl transfer. The objection to this notion is that substances such as tetramethylammonium (TMA), which have similar biological activity, cannot form such acetylated membrane intermediates. This led us to consider that TMA acts by presynaptic release of bound ACh. Thus it became necessary to devise a method which would detect ACh in the presence of TMA and like substances. The method which we have worked out depends upon the precipitation of ACh out of solution as a Reinecke complex followed by absorption to a resin. The column is treated with hydrazine which reacts 200 fold more rapidly than with non-active esters. The acetyl hydrazine is eluted and reacted with salicylaldehyde, which is employed to fluorometrically estimate the ACh content of the sample.

Method

ACh is precipitated as a Reinecke and transferred to small four-inch colunns fabricated from 8 mm Pyrex tubing which contained Bio-Rex 70 (50-100 mesh) in the sodium form. The ACh is taken up by the resin at the top of the column. Hydrazinolysis of the ACh is achieved by placing 0.1 ml of 2 M-hydrazine on the column for 5 minutes, and the acetyl hydrazide formed is eluted with 2.5 ml of a 0.07 N-HCl. The eluate is treated with 0.1 ml of 4% salicylaldehyde in dimethylformamide for 30 minutes at 37°C. The tubes are treated with 0.1 ml of 0.2 M-KBH_4 for 8 minutes at room temperature followed by 0.1 ml of 2 N-NaOH and the fluorescence read immediately in an Aminco-Bowman spectrophotofluorometer (activation 37 mμ fluorescence 475 mμ). This method can detect 0.5 mμM-ACh and a linear relationship between fluorescence and concentration is observed for up to at least 10 mμM. The specificity of this procedure has been explored in some detail and the application of this method for the study of presynaptic release of ACh will be discussed.

This work was supported by PHS grant No. NB. 1572.

DOPAMINE IN CORPUS STRIATUM AND BEHAVIOUR IN RATS

R.L. Fog, J. Scheel-Krüger, A. Randrup,
I. Munkvad and H. Pakkenberg

Sct. Hans Hospital, Department E, Roskilde and
Kommunehospitalet Laboratory of Neuropathology, Copenhagen.

Dopamine in the corpus striatum seems to play an important part in the mechanisms of stereotyped behaviour in the rat. Amphetamine given to rats induces stereotyped sniffing, licking and biting. This behaviour can be inhibited by α-methyl-tyrosine, which inhibits the synthesis of DOPA (and thus the synthesis of both dopamine and noradrenaline). Restoration of the stereotyped behaviour can be obtained by subcutaneous injections of DOPA or anticholinergic agents. In contrast, diethyldithiocarbamate (DDC), which inhibits the synthesis of noradrenaline but not that of dopamine, does not inhibit the amphetamine stereotypies.

A restoration of stereotypies is also obtained by micro-injection of dopamine in the corpus striatum in rats, in which the amphetamine-behaviour is inhibited by α-methyl-tyrosine. A more rapid restoration is obtained by simultaneous microinjection of both dopamine and an anticholinergic agent.

Subcutaneous injection of a monoamineoxidase-inhibitor and DOPA produces in rats a rage reaction followed by stereotyped sniffing, licking and biting. Pretreatment with reserpine and DDC strongly reduces the formation of noradrenaline but not that of dopamine. Behaviourally, the rage reaction is reduced, but the stereotypy remains unaffected.

Dopamine thus seems to be responsible for the stereotyped behaviour, while noradrenaline (perhaps supported by dopamine) seems to be involved in other forms of activity such as aggressive behaviour.

The role assigned to the corpus striatum agrees with other experiments in which intracerebral microinjection of various neuroleptic drugs in the corpus striatum instantaneously inhibited the amphetamine-induced stereotyped behaviour.

METABOLITES IN THE BRAIN DURING THE SEIZURES EVOKED BY METHIONINE SULPHOXIMINE

J. Folbergrova, O.H. Lowry and J.V. Passonneau

Department of Pharmacology and the Beaumont-May Institute of Neurology, Washington University School of Medicine, St. Louis, Missouri.

The levels of glucose, glycogen, ATP, P-creatine, lactate, glutamate, and ammonia were measured with the aid of enzymic fluorometric techniques in mouse brain after administration of the convulsive agent methionine sulphoximine (MSI). The effect of MSI on the quantitative histochemical distribution of glucose and glycogen was also determined in 4 layers of the cerebral cortex and subjacent white matter.

Control values (in m-moles/kg wet weight $\pm$ S.E.) were: glucose 1.12 $\pm$ 0.12, glycogen 2.10 $\pm$ 0.18, ATP 3.02 $\pm$ 0.08, P-creatine 4.42 $\pm$ 0.11, lactate 0.80 $\pm$ 0.11, glutamate 12.4 $\pm$ 0.35, NH_3 0.244 $\pm$ 0.011. During seizures (i.e. 3.8 - 5 hr after administration of MSI) there were no changes in ATP and P-creatine. However, glucose and glycogen both increased by 130%; lactate increased 65%, glutamate decreased 20%, and ammonia increased more than 4-fold. Glucose, glycogen, and ammonia were significantly increased during the preparoxysmal period, when no signs of neurotoxic effect could be observed (1,2 and 2.8 hr respectively after administration of MSI).

When seizures were completely prevented by administration of methionine at the same time as the MSI, the increase in NH_3 was reduced by 50%, but there was no reduction in the high levels of glucose and glycogen.

The histochemical studies of glucose and glycogen distribution showed differential effects of MSI on different parts of the brain. Whereas glucose was increased equally in all cortical layers studied, including subjacent white matter, glycogen was increased to the greatest degree in the outer cortical layers, whereas there was no increase at all in subjacent white matter.

The mechanism of the neurotoxic effect of MSI is not clear. The unchanged levels of ATP and P-creatine, in spite of severe convulsions, together with increased levels of glucose and glycogen, are surprising. It is unlikely that exhaustion of energy or disturbance of carbohydrate metabolism could have a causal relationship with the convulsions. On the other hand, the accumulation of ammonia, which begins in the preparoxysmal period might be related to the neurotoxic effect of MSI.

Supported in part by grants from the U.S.P.H.S. grant (NB-05221) and the American Cancer Society grant (P-38A0.

ACCUMULATION OF [^{131}I]TRIIODOTHYRONINE IN NEURONS AND IN OTHER TISSUES

Donald H. Ford

Department of Anatomy, State University of New York, Downstate Medical Center, Brooklyn, New York

Previous investigations demonstrated the accumulation in the brains of various animals of [^{131}I]triiodothyronine (T_3) which was located primarily in neuronal regions. The vascular route was selectively better than any other for ensuring the entry of labelled hormone into the brain. However, it was impossible to determine if the [^{131}I]T_3 was associated with neurons or uniformly distributed in grey matter, because excessive background fogging with [^{131}I]T_3 or the low specific activity of [^{14}C]T_3 made radioautography unsuitable. Currently developed methods for detecting radioactivity in neurons, utilizing nerve cells dissected from grey matter, now provide a convenient method for quantitatively determining ^{3}H, ^{14}C, ^{131}I or ^{35}S labelled compounds in neurons as well as in other tissues.

The following study to determine the T_3 accumulation sites in brain was undertaken. Male rats given 19.6 μg of [^{131}I]T_3 intravenously were killed by intracardiac saline perfusion at various times after injection. Spinal grey, muscle, liver, motor neuron and plasma samples were oxidized in Schöniger flasks and the ^{131}I trapped with a phenethylamine-toluene-methanol solvent. Radioactivity was detected by liquid scintillation and the data expressed as the per cent of the injected dose/g of tissue. Ascending paper chromatography demonstrated that most of the tissue radioactivity was in T_3.

Comparisons of the ^{131}I accumulation showed liver was highest, motor neurons second, spinal grey third and muscle last. Liver ^{131}I accumulated rapidly and decreased rapidly, paralleling the rapid fall in plasma ^{131}I. Muscle, spinal cord grey and neurons had slower accumulation rates which were maintained against decreasing blood hormone levels. By seven hours, motor neurons had higher ^{131}I levels than liver. The [^{131}I]T_3 accumulation in neurons suggested that there was no significant restriction in T_3 transport from brain capillaries to neurons. The low accumulation of ^{131}I observed in brain sections appear related to the failure of the tissue intermediate between neurons and capillaries to retain significant amounts of hormone, rather than to a restriction in transport or accumulation by neurons.

TOPOGRAPHIC STUDIES OF NORADRENALINE TRANSMITTER FUNCTION

Bent Formby and Jørgen Clausen

The Neurochemical Institute, Copenhagen, Denmark

Particulate fractions of different topographic areas of the central nervous system of the rat were isolated by ultracentrifugation. The fractions were shown to contain synaptic vesicles, synaptic membranes and membranes of different neuronal origin.

Although possessing certain constituents in common, the membranes showed significant differences in sialic acid free polar lipid composition.

The distribution of the specific activity of a neutral Mg^{2+}, K^+, Na^+ dependent p-nitrophenylphosphatase was determined in the particulate fractions. This enzyme was previously found to have a close relation to the transmitter function of noradrenaline.

On the basis of the data obtained, the transmitter action of noradrenaline in different areas of the rat brain is discussed.

THE EFFLUX OF POTASSIUM FROM ISOLATED SLICES OF RAT BRAIN

G. Franck*

Department of Biochemistry, University of Liège

Kinetic studies of ion exchange are so far the best means available for helping to overcome the difficulties of chemical analysis. In the case of brain slices (0.4 to 0.5 mm thick) the technique used was to load the preparation with a radionuclide during an incubation period (1 hr) followed by washing out the radioactivity (up to 4 hr). The technique used allowed us to measure the radioactivity remaining in the slice every 10 sec under various experimental conditions (metabolic impairment, electrical stimulation, etc.). In other experiments the brain was loaded with the isotope by intraperitoneal injection of an appropriate radioactive salt.

As far as potassium is concerned, the graphical analysis showed that we are dealing with a system of four compartments in series. The respective half-renewal times were 35 sec, 5 to 6 min, 15 to 17 min and 50 to 57 min. The kinetic changes obtained under the conditions of electrical stimulation (5V; 50 sec) in the presence of ouabain (10^{-4} M) or during anoxia (N_2) showed that, except for the first compartment exchanging rapidly (35 sec), we are dealing with metabolically dependent exchanges. If one compares the results of the kinetic analysis with those obtained chemically, a good agreement is found, suggesting that we are dealing with a heterogenous population of cells exchanging potassium at three different rates. Additional experimental evidence will be presented to substantiate this view and to support the conclusion arising from our chemical analysis according to which, the incubated cells reach a new metabolically dependent steady state.

*The results presented have been obtained in collaboration with Drs. M. Cornette and E. Schoffeniels.

DISTRIBUTION QUANTITATIVE DES DIVERS PHOSPHOLIPIDES DANS LES NEURONES ET LES CELLULES GLIALES ISOLES DU CORTEX CEREBRAL DE RAT ADULTE

L. Freysz, R. Bieth et P. Mandel

Centre de Neurochimie, Faculté de Médecine, Strasbourg

Dans le but d'obtenir une quantité de cellules suffisante pour permettre une étude de la répartition des phospholipides, nous avons mis au point un procédé de séparation de neurones et de cellules gliales à partir de la matiere grise du cortex cérébral de rat adulte. Après fixation du cortex par le mélange glycérol-acétone-eau (1:1:1) dans le but d'empêcher des dégradations ultérieures, les cellules sont séparées par broyage à travers un tamis de 82 µ dans un mélange sucrose 0.25 M-glycérol, suivi d'une filtration de la suspension à travers un tamis de 58 µ. Les deux types de cellules sont ensuite séparés par ultracentrifugation différentielle à l'aide de gradients de sucrose et de ficoll permettant d'isoler successivement les neurones et les cellules gliales. Chaque fraction renferme 80 p.100 de cellules du même type. La majeure partie des cellules reste intacte et présente encore des expansions.

L'étude de la répartition des divers phospholipides a été effectuée après extraction par le mélange chloroforme-méthanol 2:1 par fractionnement chromatographique sur colonne de silice et sur papier formolé. Dans chaque type de cellules nous avons mis en évidence l'existence de phosphatidyl-éthanolamines, sérines et cholines ainsi que des plasmalogènes correspondants, de sphingomyélines, de monophosphoinositides, de cardiolipides et d'éther-phosphatides. Les quantités absolues des divers phospholipides évaluées par cellule sont environ 5,6 fois plus élevées dans les cellules gliales que dans les neurones. Cependant, le taux de chaque phosphatide par rapport au P lipidique total est similaire dans les deux types de cellules. Ceci parle en faveur de l'existence de formations membraneuses plus importantes dans les cellules gliales, mais également d'une analogie structurelle de ces formations dans les deux types de cellules, donc d'un 'modèle membraneux phospholipidique' identique.

CATECHOLAMINES IN THE COCKROACH BRAIN. A FLUORESCENCE AND ELECTRON MICROSCOPE STUDY.

Nora Frontali

Istituto Superiore di Sanità, Rome.

Studies of the mushroom bodies of the brain of the cockroach (*Periplaneta americana*), by a fluorescence histochemical method, showed the presence of catecholamines in certain zones of the neuropile. The cell bodies were not fluorescent; the positive neuropile zones must thus be sites of synapses with catecholamine-containing fibres from outside the mushroom bodies. Prolonged treatment with reserpine was followed by complete disappearance of the fluorescence reaction.

The mushroom body neuropile, observed with the electron microscope, is formed almost exclusively of nerve endings. These contain mitochondria, synaptic vesicles, and a varying quantity of electron-dense granules of approximately 1000 Å diameter. The distribution of the fluorescence reaction did not parallel the distribution of the granules. Treatment with reserpine was not followed by depletion or disappearance of the granules. It is concluded that the presence of these granules is not directly linked to the storage and release of catecholamines.

Chemical analysis of the catecholamines in the cockroach brain is in progress.

NONHYDROXY AND HYDROXY FATTY ACIDS IN BRAIN CEREBROSIDES OF MAN AND VARIOUS ANIMAL SPECIES

R. Fumagalli, C. Galli and P. Capella

Department of Pharmacology, University of Milan, Italy and Department of Agriculture Chemistry, University of Bologna, Italy.

The fatty acid composition of the brain cerebrosides, which are known to contain both hydroxy and nonhydroxy fatty acids, was studied in man, bovine species, sheep and rat, as well as in different parts of the nervous systems of various species of marine fish (olfactory lobe, spinal cord, cerebellum and brain), with the object of following the variations of hydroxy fatty acids in phylogenesis.

Analyses were carried out using thin layer chromatography and gas-liquid chromatography (GLC). Structural investigations of hydroxy fatty acids were carried out using a combination of GLC and mass spectrometry. For quantitation and identification purposes, derivatives of hydroxyl substituted fatty acids (such as trimethylsilyl ethers) were used.

The variations of fatty acid composition in the species considered, with respect to their pattern and the relative content of hydroxy and nonhydroxy fatty acids, are discussed.

EFFECT OF DRUGS AND DIFFERENT EXPERIMENTAL CONDITIONS ON THE ACTIVITY OF CENTRAL MONOAMINE NEURONS

K. Fuxe, H. Corrodi and T. Hökfelt

Department of Histology,
Karolinska Institute, Stockholm

The rate of depletion of amines after inhibition of synthesis is highly dependent on the nervous impulse flow. Changes in the rate of amine depletion after synthesis-inhibition may therefore indicate changes in the activity of central monoamine neurons. A combined histochemical and biochemical analysis of dopamine (DA), noradrenaline (NA) and 5-hydroxytryptamine (5-HT) was made on rat brain after treatment with a potent inhibitor of the tyrosine hydroxylase (α-methyl-tyrosine; H 44/68) or an inhibitor of both the tyrosine and tryptophane hydroxylase (α-propyldopacetamide; H 22/54) in combination with various drugs or under certain experimental conditions. Barbiturates were found to retard the depletion in the central DA neurons obtained after H 44/68; chlorpromazine and haloperidol increased the rate of amine depletion in the DA and especially the NA neurons obtained after H 44/68; α- and β- adrenergic blocking agents increased the rate of depletion in the NA neurons. No increased rate of depletion was obtained with these drugs in NA nerve terminals lacking nervous impulse flow. Heat increased the rate of amine depletion in the 5-HT and NA neurons obtained with H 22/54. Cold retarded the amine depletion in the central 5-HT neurons obtained after H 22/54. The results strongly indicate that barbiturates decrease directly or indirectly the activity of the central NA neurons, whereas ethanol and α- and β-adrenergic agents increase the activity in the central NA neurons. Chlorpromazine and haloperidol probably increase the activity in both the central DA and NA neurons. The results furthermore underline the view that central NA- and 5-HT neurons participate in thermoregulation.

QUANTITATIVE ANALYSIS OF INDIVIDUAL MOLECULAR SPECIES OF LIPIDS IN THE NERVOUS STRUCTURES OF MARINE ANIMALS

Claudio Galli

Institute of Pharmacology, University of Milan, Milan, Italy.

Total lipids, sterols, cerebrosides and individual phospholipids were analysed in various nervous structures (brain, cerebellum, olfactory bulb) of several marine animal species including Elasmobranches, such as sharks (Orectolobidae, Ginglymostoma cirratum and Carcharinidae Negaprion Brevirostris), sting rays (Dasyatidae, Dasyatis Americana) and fishes such as skip jacks (Carangidae, Caranx Ruber).

Sterols and cerebrosides were determined using monodimensional thin layer chromatography (TLC), while phospholipids were quantitated with phosphorus assays on two-dimensional TLC. The values were compared with mammalian (rat, beef, human) brain lipid composition.

The results indicate that brain lipid composition in the various animal species is qualitatively similar, but there are significant quantitative differences. In Elasmobranches, brain, cerebellum and olfactory bulb showed different total lipid concentration and phospholipid composition. Brain had the highest cerebroside and sterol concentration, while cerebellum and olfactory bulb had the highest lecithin content. The fish (skip jack (Carangidae, Caranx Ruber) had a brain lipid composition similar to cerebellum in elasmobranches.

Similarities and differences in relation to phylogenetic evolution of central nervous tissues are discussed.

BRAIN STEROLS IN THE ADULT RAT

G. Galli* and E. Grossi Paoletti

Institute of Pharmacology, University of Milan.

Cholesterol is generally considered to be the only sterol in adult nervous tissue, while desmosterol is also present in immature brain (PAOLETTI *et al.*, 1965). Trace amounts of other sterols have been detected in adult brain in various laboratories independently, as reviewed by ROSSITER (1962). Active cholesterol biosynthesis occurs in developing brain, but it is retained only at low levels in the adult (DAVISON, 1965). The possibility exists that part of the radioactivity found in adult brain cholesterol after treatment with labelled metabolites is associated with very small, undetectable pools of other sterols. Their identification is particularly difficult because of methodological problems.

A column chromatographic method, using silver nitrate-silica gel G:celite as adsorbent, was developed for the separation of brain sterols (GALLI & GROSSI PAOLETTI, 1967).

Brain sterols were obtained from rats injected with [2-^{14}C]mevalonic acid and sacrificed one hour later. The sterols were acetylated and fractionated on the column. Cholesterol was the major component, but other sterols, accounting for no more than 2% of the original mixture, were identified and quantitatively measured by gas chromatography. The molecular weights and structures of these sterols were determined with the aid of gas chromatography-mass spectrometry. The distribution of the radioactivity in the various sterols is discussed.

*Fellowship from the Commission for Scientific Research of Italian Switzerland

DAVISON, A.T.N (1965). In *Advances in Lipid Res.*, 3, 171, Academic Press, N.Y. GALLI, G. & GROSSI-PAOLETTI, E. (1967). *Lipids*, 2, 36. PAOLETTI, R., FUMAGALLI, R., GROSSI, E. & PAOLETTI, P. (1965). *J. Am. Oil Chem. Soc.*, 42, 400. ROSSITER, R. (1962). In *Neurochemistry*. Edited by K.A.C. Elliott, I.H. Page & J.H. Quastel, p.10. C. Thomas, Springfield, Ill.

TECHNICAL PROCEDURE FOR MICRO DISC-ELECTROPHORESIS

S. Gandini and A.R. Dravid

Neuro-psychiatric Clinic, University of Pisa, Pisa, Italy, and Institute of Neurobiology, University of Göteborg, Sweden.

Some basic conditions necessary to adapt acrylamide disc-electrophoresis to a microscale were considered from a thermodynamic point of view, and a general formula was developed by taking into account the main factors interfering with the system. Starting from the statement that the intensity of the electrical current should be the highest to achieve the best electrophoretic resolution, the maximal level of heating inside the gel was preset at the stationary state. The assumption was made that some coefficients are known and constant: thermal and electrical conductivity of the gel, and variation of electrical resistance of the gel with temperature. It was then possible to deduce from the general formula that an optimal relationship links the diameter and the thickness of the glass tube, the coefficient of thermal conductivity of the glass and the thermal-exchange coefficient between glass and external cooling fluid.

A further preliminary condition is that the temperature gradient inside the gel should not exceed a critical value which is a function of the diameter, in order to avoid a dome-like bending of the bands.

On the basis of these statements it was possible to define the whole system, by calculating the optimal values for diameter, thickness and thermal conductivity of the glass tube, with different external cooling systems (air in motion; cold running water). At the end of the 'steady-state stacking' the capillary tube was cut just beyond the starting point in order to decrease the resistance and to obtain the maximum amount of current for electrophoretic separation of the proteins. A special power generator was designed with five independent channels giving an intensity of current in a range between 0 and 100 μAmps each channel. Finally, a microdensitometric procedure is described, based on the scanning of photographic plates on a Joyce-Loebl apparatus. Electrophoretic separation of very small amounts of soluble proteins, corresponding to the protein content of a pool of single nerve cells, with a high degree of resolution, is made possible by the present procedure.

EFFECTS OF SEVERE PROTEIN-CALORIE DEFICIENCY ON THE INCORPORATION OF GLUCOSE CARBON INTO FREE AMINO ACIDS OF RAT BRAIN CORTEX IN VIVO

Jacques Gayet and Paul Lehr

Laboratoire de Physiologie Générale,
Université de Nancy, France.

Measurements were made of the rate of incorporation of ^{14}C, after subcutaneous injection of [U-^{14}C]glucose in the adult rat, into free aspartic, glutamic and γ-aminobutyric acids and free neutral and basic amino acids of the brain cortex. The animals were given the special diets previously described (LEHR & GAYET, 1963).

(1) In rats eating a complete purified diet *ad libitum*, at 30 minutes after injection the distribution of ^{14}C in the free amino acids of the cerebral cortex was: aspartate 11 per cent, glutamate 54 per cent, γ-aminobutyrate 9 per cent, neutral 25 per cent and basic 1 per cent. The specific radioactivity of γ-aminobutyrate was about the same as, or higher than, that of glutamate. These results are in accordance with those previously published by GAITONDE *et al.* (1965).

(2) In rats submitted to 30 days or 60 days deprivation of dietary protein, at 30 minutes after injection the distribution of ^{14}C in free amino acids of the cortex was about the same as in the cortex of rats given a complete purified diet *ad libitum*: but the specific radioactivity of aspartate was higher than that of γ-aminobutyrate and glutamate.

(3) In rats eating for 53 days a complete purified diet, equivalent in amount to that eaten daily by the rats submitted to protein deprivation, at 15 minutes after injection the distribution of ^{14}C in free amino acids of the cortex was: aspartate 11 per cent, glutamate 54 per cent, γ-aminobutyrate 9 per cent, neutral 24 per cent and basic 1 per cent. The specific radioactivity of aspartate was higher than that of γ-aminobutyrate and glutamate.

These results are discussed according to the hypothesis previously put forward (LEHR & GAYET, 1966).

GAITONDE, M.K., DAHL, D.R. & ELLIOTT, K.A.C. (1965). *Biochem. J.*, **94**, 345. LEHR, P. & GAYET, J. (1963). *J. Neurochem.*, **10**, 169. LEHR, P. & GAYET, J. (1966). *J. Neurochem.*, **13**, 805.

PHENYLALANINE AMINOTRANSFERASE. PROPERTIES AND FUNCTIONS

Harvey George and Sabit Gabay

Biochemical Research Laboratory, Veterans Administration Hospital, Brockton, Mass., and Department of Biochemistry, Boston University School of Medicine, Boston, Mass.

Pig brain phenylalanine oxoglutarate aminotransferase was purified approximately 500-fold by a procedure involving sonication, $MnCl_2$ precipitation, $(NH_4)_2SO_4$ fractionation, ultrafiltration, and chromatography on DEAE-cellulose. The assay used throughout the experiments herein described consisted of two procedures devised in this laboratory. One of these, a spectrophotometric assay employed a borate-arsenate enolizing mixture; the other, a radiometric procedure, made use of ion exchange paper and permitted a direct determination of the per cent of substrate converted to product. On polyacrylamide gel electrophoresis, the enzyme appeared to be made up of three slow-moving components. The K_m values for L-phenylalanine and α-ketoglutarate (α-KG) were 5×10^{-2} M and 6×10^{-4} M, respectively. High concentrations of α-KG were inhibitory; half maximal inhibition was observed with 5×10^{-2} M α-KG in the presence of 15 μmoles phenylalanine. The inhibition of the transaminase by α-KG was dependent upon phenylalanine concentration in the tested range of 10^{-2} to 10^{-1} M. Pyridoxal phosphate (PAL) acted both as a coenzyme and as an activator. The enzyme exhibited no initial lag period, but only half of its normal activity was observed when PAL was omitted from the reaction mixture. The apoenzyme (formed by 5 min. 50° preincubation in 0.5 M-phosphate buffer, pH 5.7) was inactive and could be rapidly reactivated when preincubated with its coenzyme. However, when the holoenzyme was preincubated with additional PAL, prior to the addition of substrate, its activity was markedly increased. Metal ion chelators (10^{-2} M-EDTA, 10^{-2} M-Cupferron, 10^{-3} M-sodium diethyldithiocarbamate) had no effect on enzymic activity, nor did high concentrations of ammonium sulphate (1.0 M). The enzyme was inhibited by p-chloromercuribenzoate and inhibition by such sulfhydryl characterizing reagents could be prevented by the presence of substrate.

The effect of various concentrations of phosphate, arsenate, cacodylate, borate and tris buffers on the reaction were investigated. Under the conditions employed phosphate and arsenate were appreciably (and equally) active, while much lower activity was observed with cacodylate, borate and tris. No activity was observed with glycine buffers.

In the light of these observed properties, the regulatory function of phenyl aminotransferase in the nervous tissue and its role as precursor of biologically active phenylamines were evaluated.

BIOCHEMICAL AND BIOPHYSICAL CHARACTERIZATION OF CYTOPLASMIC AND NUCLEAR RIBONUCLEOPROTEINS AND RIBONUCLEIC ACIDS OF RAT BRAIN CORTEX

J.J. Ghosh

Biochemistry Department, Calcutta University, Calcutta-9.

Methods have been described for the preparation of cytoplasmic and nuclear ribosomes, as far as possible free from heavier polyribosome fractions, from rat brain cortex tissues. The purity of rat brain cortex cytoplasmic and nuclear ribosomes has been established on the basis of (a) constancy of protein and RNA content, (b) absence of certain marker constituents and enzymes, (c) ultraviolet absorption spectral characteristics (d) sedimentation analysis, (e) intrinsic viscosity determination and (f) molecular weight determinations.

The rat brain cortex cytoplasmic ribosomes used in this study, had a major component with a sedimentation coefficient of 82.4 S at 20° C and infinite dilution in sucrose-KCl-Mg-cacodylate (SKM) buffer. In the absence of Mg^{++} ion, the major component dissociated into smaller components viz., 62.5 S and 38.2 S at infinite dilution and at 20° C. The intrinsic viscosity of the 82.4 S component was found to be between 0.055 dl/g and 0.068 dl/g. A combination of sedimentation coefficients and intrinsic viscosity values was used to calculate the molecular weight of the 82.4 S component, which was found to be about $4.9 \times 10^6 \pm 15\%$.

Sedimentation analysis of nuclear ribosomal preparations showed the presence of a major component with a sedimentation coefficient of 78.5 S (corrected to infinite dilution) in SKM buffer. In 0.25 M-sucrose (in absence of Mg^{++} and K^+ ions) two other slower moving components viz., 58.2 S and 40.8 S appeared. The intrinsic viscosity of the 78.5 S component of nuclear ribosomes was found to be between 0.068 to 0.080 dl/g. The molecular weight of the 78.5 S component, calculated from the sedimentation viscosity data, was found to be about $4.2 \times 10^6 \pm 15\%$.

Cytoplasmic ribosomal RNA, used in the present study, consisted of two nearly homogeneous components, viz., 31.4 S and 18.2 S (at infinite dilutions and at 20° C) in 0.01 M-Tris buffer (pH 7.4) and 10^{-4} M-$MgCl_2$. Intrinsic viscosity values of the 31.4 S and 18.2 S components, after separation by sucrose density gradient centrifugation, were found to lie between 0.32 to 0.37 dl/g and 0.20 to 0.30 dl/g respectively. The molecular weights of 31.4 S and 18.2 S, calculated on the basis of sedimentation-viscosity data, were 1.7×10^6 and 5.2×10^5 respectively.

Nuclear ribosomal RNA preparations consisted of two nearly homogeneous components viz., 28.2 S and 15.3 S, corrected to infinite dilutions and at 20° C. Intrinsic viscosity determinations of 28.2 S and 15.3 S components of nuclear ribosomal RNA were made after separation of these two components by sucrose gradient centrifugation. It was found that the intrinsic viscosity values of 28.2 S and 15.3 S were in the range of 0.42 to 0.52 dl/g and 0.32 to 0.38 dl/g respectively. The molecular weights of 28.2 S and 15.3 S components of nuclear ribosomal RNA were 1.4×10^6 and 4.8×10^5 respectively.

ACTION OF TRANQUILLIZER DRUGS AT THE LEVEL OF BRAIN NUCLEOPROTEINS

J.J. Ghosh, Swati Ghosh, S. Chanda, K. Sikdar and S. Bhaduri

Department of Biochemistry, Calcutta University, Calcutta-9

In recent years considerable interest has developed in drug-induced alterations in the nucleic acid constituents of nerve cells resulting in structural and functional impairment of the central nervous system (GHOSH, DATTA & BHATTACHARYYA, 1965; SIKDAR & GHOSH, 1964; KOENIG, 1960; HYDÉN, 1962). In an attempt to elucidate the action of the phenothiazine group of drugs at the level of the structural constituents of nerve cell bodies, it was found that treatment of rats with different doses of chlorpromazine, prochlorperazine and chlordiazepoxide produced nuclear swelling and distinct chromatolytic changes in the Nissl granules of nerve cells in different areas of the rat brain. Among the different areas studied, thalamo-hypothalamic regions showed the earliest onset of structural changes at the nuclear and cytoplasmic levels, with subsequent restitution. Next in order of structural changes were the nerve cell bodies in brain stem regions, which were followed by motoneurons of the spinal cord.

Further studies made to interpret the chromatolytic changes in terms of the different factors affecting the stability of cellular ribonucleoproteins, indicate that both enzymic and non-enzymic factors are involved in drug-induced chromatolytic changes in the cytoplasmic Nissl granules. Among the different non-enzymic factors, changes in the concentrations of monobasic or dibasic cations like K^+, Na^+ or Mg^{++} ions, weakening of intramolecular hydrogen bonds and/or changes in pH surrounding the cytoplasmic ribonucleoproteins are possibly involved during the action of these drugs. The significant increase in both nuclear and nucleolar sizes is also accompanied by a dissociation or release of lysine-rich histone components from the nuclear regions in the motoneurons of drug-treated rats.

NEW MICROCHEMICAL APPROACHES TO THE STUDY OF THE FUNCTIONAL ACTIVITY OF INDIVIDUAL NEURONS

Ezio Giacobini

Department of Pharmacology, Karolinska Institutet, Stockholm

The microchemical methods presently available offer sufficient sensitivity for accurate analysis of individual neurons. It remains, however, to be seen whether such a sensitivity is high enough to detect biochemical changes occurring during functional activity. In order to tackle this problem, different micromethods (the Cartesian diver, microfluorimetric techniques, isotopic microtechniques and microflamephotometry) have been combined. In some cases they have also been applied in close connection with neurophysiological methods in a study performed on isolated and functional invertebrate neurons.

The results of the following studies will be reported:

(a) variation of oxidative activity in neurons at rest and during functional activity;

(b) effect of metabolic and ion transport inhibitors on the electrical activity and oxygen consumption of isolated invertebrate neurons;

(c) variations of metabolites, energy rich phosphates and pyridine nucleotides after prolonged impulse activity;

(d) effect of anoxia on pyramidal cells from the hippocampus of vertebrates;

(e) identification of neuronal populations within the sympathetic system with respect to their adrenergic or cholinergic function.

The results obtained from the studies of points (a), (b) and (c) show that a number of impulses as low as 50,000-100,000 initiates in the neuron a sequence of biochemical changes detectable as variations in oxygen consumption and levels of metabolites. After 120,000-240,000 impulses the levels of these metabolites have reached new values. An implication of glycolytic processes in the mechanism leading to impulse activity is suggested.

Concerning point (e), using a new isotopic microtechnique, it was possible to outline different cell populations in the sympathetic ganglia showing specific differences in enzymic activity. In this study, acetylcholinesterase, cholinacetylase, and monoaminoxidase activity was determined. The results suggest a remarkable difference in the number of cholinergic, resp. adrenergic neurons in different ganglia and also indicate the possibility that the same neuron can metabolize two different chemical transmitters.

EFFECTS OF VARIOUS FORMS OF STRESS ON THE TURNOVER OF [^{3}H]NOREPINEPHRINE IN THE RAT BRAIN

J. Glowinski*, A.M. Thierry, F. Javoy and S.S. Kety

National Institute of Mental Health, Bethesda, Maryland

Considerable interest attaches to the possible role of norepinephrine and other biogenic amines in various parts of the nervous system in affective states. The evidence for such roles has been fairly indirect, depending upon an interpretation of the actions of certain psychoactive drugs or the measurement of norepinephrine levels in brain tissue under various forms of stress, with or without blockade of its synthesis by α-methyltyrosine. In the present studies information on turnover of norepinephrine in various regions of the brain was obtained by examination of the disappearance of [^{3}H]norepinephrine injected intraventricularly or intracisternally. In some experiments [^{3}H]dopamine was injected and the endogenously produced [^{3}H]norepinephrine examined. The forms of stress included repetitive electroshock to the feet, exposure to cold, rigorous confinement, and the conditioned emotional response as a type of 'psychological' stress unaccompanied by noxious physical input. In general the disappearance of the radioactive norepinephrine was considerably greater than the decrease in the level of the endogenous amine, indicating a greater effect of these stresses on turnover than on simple release of the amine and the greater sensitivity of such measurements in comparison with measurements of the level of endogenous norepinephrine. Physical stress appeared to have a significantly greater effect than psychological stress on norepinephrine turnover in the brain. Attempts were made to elucidate the pathways mediating these changes by making appropriate stereotactically placed lesions.

(* Collège de France, Paris 5ème)

INFLUENCE OF THE SUBSTANTIA NIGRA ON THE SYNTHESIS OF BIOGENIC AMINES IN THE STRIATUM

M. Goldstein, B. Anagnoste, W.S. Owen and A.F. Battista

Departments of Psychiatry and Neurosurgery
New York University Medical Center
Neurochemistry Laboratory,
New York.

Unilateral radiofrequency lesions were placed in the left ventromedial tegmental region of the rostral pons of the monkeys. Tyrosine hydroxylase activity was determined one to six months later in the striatum on the lesion side and on the intact side. The enzymic activity was found to be significantly lower on the lesion side of the caudate nucleus and putamen. Also, the _in vivo_ formation of dopamine-^{14}C from [^{14}C]tyrosine was impaired on the lesion side in these two regions of the CNS. The effects of these lesions on the biosynthesis of serotonin and other biogenic amines are now under investigation. In some other experiments unilateral lesions were placed in the lateral basal ventral thalamus. Preliminary results show that the endogenous dopamine levels are decreased on the lesion side in the striatum but the storage of [^{3}H]dopamine is not impaired. It is conceivable that lesions in the VPL region produce a transynaptic effect.

(Supported by USPHS Grants MH 02717, NB 5480 and NB 04257 and by NSF Grant GB 3176).

HETEROGENEITY OF BEEF BRAIN S-100 PROTEIN

G. Gombos, K. Uyemura, J. Tardy and G. Vincendon

Centre de Neurochimie du C.N.R.S., Faculté de Médecine, Strasbourg.

The S-100 protein, described by B.W. MOORE as specific of the nervous system, was isolated by a simplified technique. The protein prepared by our method and that purified by the MOORE technique had identical electrophoretic migration, the same $S_{w,20}$ in the ultracentrifugal analysis and both reacted against the S-100 antiserum.

The S-100 protein when submitted to electrophoresis on polyacrylamide gels or polyacrylamide-agarose mixed gels, showed a considerable electrophoretic heterogeneity. The number of the bands depended upon the buffer and the concentration of acrylamide used. Several bands were eluted separately and submitted again to electrophoresis. Each of them migrated again as an unique band.

These results and the fact that we observed several N-terminal amino acids, seem to indicate that the S-100 protein, hitherto considered as a single protein, actually is a mixture of proteins. The ultracentrifugal analysis of S-100 protein fraction in the different buffers used for electrophoresis was carried out. The protein fraction has a $S_{w,20}$ of 2.03 ± 0.09. The Schlieren patterns showed a single symmetrical apparently monodisperse, fast diffusing peak. This result indicates that the various components of the S-100 protein fraction have similar shape and molecular weight.

CHANGES IN NUCLEIC ACIDS AND ENZYMES IN DEVELOPING CEREBRAL CORTEX AND CEREBELLUM OF THE RAT AFTER NEOTAL THYROIDECTOMY

Carlos J. Gómez, Juana M. Pasquini and Carlos Gargia Argiz

Departamento de Química Biológica, Facultad de Farmacia y Bioquímica, Universidad de Buenos Aires

The changes in nucleic acids, succinate dehydrogenase (SDH), glutamate decarboxylase (GAD), GABA transaminase (GABAT), aspartate amino transferase (AAT) and adenosin-triphosphatase (ATPase) were studied in developing cerebral cortex and cerebellum from normal and neonatally thyroidectomized rats, during the first 50 postnatal days.

The values obtained for RNA and enzymic activity, expressed as a function of wet tissue weight, were compared with values expressed on a DNA basis. The latter appears to give a better indication of the biochemical and morphological changes related to maturation.

During normal postnatal morphogenesis, with the exception of the nucleic acids, the patterns of change were similar in the cerebral cortex and the cerebellum. Neonatal thyroidectomy produced an increase in the DNA content of the cerebral cortex, but only slightly affected that of the cerebellum.

With the exception of the soluble AAT, the lack of thyroid function produced in the cerebral cortex a substantial decrease in the activity of all the enzymes studied. In the cerebellum it also led to a significant decrease in the activities of mitochondrial AAT, GABAT and $Na^{+}K^{+}$-ATPase; but SDH and GAD were only temporarily affected and Mg^{++}-ATPase was unaffected. The significance of these results is discussed in connection with the function of thyroid hormones on protein synthesis in immature brain.

ROLE OF LIGHT IN THE BIOSYNTHESIS OF RIBONUCLEIC ACIDS DURING ONTOGENESIS OF VISUAL PATHWAYS

S. Goswamy and P. Mandel

Centre de Neurochimie du C.N.R.S., and Institut de Chimie Biologique, Faculté de Médecine, Strasbourg.

The role of light in RNA biosynthesis, during ontogenesis of the retina and visual cortex of the rabbit has been explored. We have previously reported a retardation in the development of the pattern of distribution of proteins and nucleic acids in the visual pathways of young rabbits deprived of visual stimuli since birth. Employing ^{32}P as precursor and taking into account the biosynthesis of the free nucleotides, we observed that during the course of normal development of the retina and visual cortex, there is an increase of RNA synthesis from birth to 21 days. The quantities of free nucleotides, specially of the adenylic type, rise considerably in this period, which may be related to increased functional activity. A histoautoradiographic study revealed three periods of retinal maturation: (1) before the opening of the eyes, having a high rate of RNA turnover; (2) about the time of opening of the eyes, with lower RNA turnover; (3) after opening of the eyes and related to the functional response to visual stimuli, accompanied by a progressive increase of RNA synthesis up to about 21 days. During the third period, the RNA synthesis was further increased by stimulation with light.

In the animals brought up in total darkness for 21 days since birth, the conversion of nucleoside triphosphates into RNA is reduced in the absence of light. The [^{3}H]uridine incorporation in the cellular layers of the retina during the functional maturation period (from 8 to 21 days after birth) in animals deprived of light since birth reflected a retardation in the evolution of RNA synthesis, in the absence of functional stimuli. These changes are most pronounced in the outer nuclear layer of the retina.

TRANSIENT LOSSES OF CELL LIPID FROM CULTURES OF HUMAN FOETAL CEREBELLA EXPOSED TO HETEROLOGOUS SERUM

F. Güttler and J. Clausen

The Neurochemical Institute, Copenhagen.

Reversible changes in the lipid composition of lung tissue following anaphylaxis has recently been reported (GOADBY & SMITH, 1966). Cultures of nervous tissue were examined in order to investigate discrete changes in the lipid composition of cells maintained in a well-defined system. Cultures of human foetal cerebella grown in a synthetic medium supplemented with 40% serum exhibited a healthy population of astrocytes, oligodendrocytes, microglial cells and neurons (Purkinje cells and Golgi cells), some with myelinated axons. Pairs of 75-day old cultures fixed a certain time intervals were used to investigate possible cytopathic injury following application of heterologous serum, i.e. freshly obtained normal rabbit serum. The cultures were carefully washed and sacrificed by immediate transfer to -85°C methanol. The cytopathic effect was tested by measuring the release of cellular RNA. The DNA content of the freeze-dried cultures was estimated fluorometrically and the lipid phosphorus, cerebrosides and cholesterol content of chloroform-methanol extracts were expressed per μg DNA. Polar lipids were estimated by adaption on a micro-scale of the method described by CLAUSEN _et al_. (1965).

The RNA content of the media increased 22% during the first two hours of incubation, (expressed by the cytopathic index of ALLAN & MALKINSON, 1966). The loss of RNA fell slightly within the next 20 hr. Substantial changes occurred in the phospholipids. Lipid phosphorus fell 85% in 35 min. but had returned to normal within 23 hr. This was accompanied by an 88% loss in cholesterol and cerebrosides in 20 min and 3 hr., respectively. Cholesterol had returned to normal within 4 hr. and cerebrosides within 23 hr.

ALLAN & MALKINSON (1966). _Nature, Lond._, _211_, 493. CLAUSEN _et al_. (1965). _J. Neurochem._, _12_, 599. GOADBY & SMITH (1966). _Biochem. J._, _100_, 30P.

SOME STUDIES ON THE EFFECT OF ELECTRICAL STIMULATION ON THE SYMPATHETIC ADRENERGIC NEURON

Jan Häggendal

Department of Pharmacology,
University of Göteborg, Göteborg, Sweden.

Data have recently been obtained permitting calculations on the amount of endogenous noradrenaline (NA) released during nervous activity from the nerve terminals of peripheral adrenergic neurons (FOLKOW & HAGGENDAL, 1967). The amount of NA was estimated in the arterial blood and in the venous effluent from skeletal muscles of the hind limb in the cat. The adrenergic nerves to this area were stimulated electrically. The blood flow was measured and could be changed by stimulation of the motor nerves to the muscles.

The amount of NA in the venous effluent was dependent on several factors, such as the blood flow and the re-uptake of released NA back into the varicosities of the nerve terminals by the so-called 'membrane pump'. When the blood flow was large and substances blocking the 'membrane pump' had been given, 1/25000 of the tissue content of NA was obtained in the venous effluent per electrical impulse, if the frequency of stimulation was 6 imp. per second. Data from studies on the intraneuronal distribution of NA (DAHLSTROM & HAGGENDAL, 1966; DAHLSTROM, HAGGENDAL & HOKFELT, 1966) were used to calculate the approximate amount of NA released from one varicosity of the nerve terminals per impulse. The amount of released NA was thus found to be about 1/25000 of the NA content of one average varicosity, which may contain about 1500 amine storage granules.

These results, together with those presented here by Dr. DAHLSTROM, make it unlikely that whole NA storage granules would be released during normal nervous activity. However, during high and prolonged nervous activity of these neurons it seems probable that besides an increased release of NA there also occurs an increased down-transport of newly-synthesized granules from the cell bodies (DAHLSTROM & HAGGENDAL, 1967). The increased activity would thus also imply an increased protein synthesis, since the storage granules are themselves complex protein structures. Preliminary results will be given and discussed with respect to this question.

BIOSYNTHESIS OF GLYCOLIPIDS: INCORPORATION OF RADIOACTIVE N-ACETYL GALACTOSAMINE FROM URIDINE-DIPHOSPHO-(N-ACETYL) [^{14}C]HEXOSAMINE INTO GLYCOLIPIDS BY A RAT BRAIN PARTICULATE PREPARATION

Shizuo Handa and Robert M. Burton

Department of Pharmacology and Beaumont-May Institute of Neurology, Washington University School of Medicine, St. Louis, Missouri

Incorporation of [1-^{14}C]glucosamine into rat brain gangliosides, *in vivo*, has been shown (BURTON *et al*, 1963). Assuming step-wise synthesis, we searched for UDP-(N-acetyl) galactosamine: sphingolipid transferase activity in rat brain.

A mixture (7-3) of UDP-(N-acetyl)[1-^{14}C]glucosamine and UDP-(N-acetyl)[1-^{14}C]galactosamine was prepared by *in vivo* procedures. Using the UDP-(N-acetyl)[1-^{14}C]hexosamine mixture as a substrate, a particulate enzyme preparation was demonstrated in young rat brain tissue which catalyzes the reactions:

(1) UDP-glcNAc ⇄ UDP-galNAc (E.C.5.1.3)

(2) UDP-galNAc + gal-glc-cer ⟶ galNAc-gal-glc-cer + UDP

The glycolipid acceptor specificity was examined and preliminary kinetic data obtained for the UDP-(N-acetyl)-galactosamine: gal-glc-ceramide N-acetyl galactosamine transferase. The products of the reaction were identified. The relationship of this reaction to ganglioside biosynthesis will be discussed.

(Supported in part by USPHS grant NB-01575-09).

BURTON, *et al*. (1963). *Biochem*. **2**, 580.

AN INTERPRETATION OF NEURONAL FUNCTION, CYTOCHROME OXIDASE ACTIVITY AND TRICYANOAMINOPROPENE

Joseph Harris and Kim Ritchie

Barrow Neurological Institute, Phoenix, Arizona and Arizona State University, Tempe, Arizona.

Functional changes in the nervous system are considered to be reflected in the level of cytochrome oxidase activity. HYDEN (1961) and his group have reported that neuronal stimulation results in increased cytochrome oxidase activity together with increased protein and RNA formation. A parallel result was reported by them by the action of tricyanoaminopropene (TRIAP), a compound they found capable of decreasing learning time. Their interpretation is that, "TRIAP stimulates RNA and consequently the protein production and increases the energy utilization in the neurons."

This interpretation, especially regarding cytochrome oxidase activity, may bear reconsideration in light of our data on the action of TRIAP on cytochrome C. We have found that, aerobically or anaerobically, a chemical reduction of cytochrome C is slowly brought about by TRIAP; the rate of reduction is markedly increased by the presence of the electron acceptor 2,6-dichlorophenolindophenol (DCPIP). The reduced cytochrome C, the normal substrate for cytochrome oxidase, will undergo aerobic reoxidation by TRIAP in the presence of Cu^{++} or Cu^{-} protein, e.g., ceruplasmin. This then constitutes an *in vitro* model of cytochrome oxidase in which TRIAP participates in cytochrome C oxidation at rates characteristic of the complete system *in vivo*. This model suggests a possible non-enzymic basis for the increased cytochrome oxidase activity induced by TRIAP as reported by HYDEN. Confirmation of this is currently under study in the intact animal. Nerve cells are considered to achieve priority for the available energy of the respiratory chain whenever brain activity increases. Therefore the control mechanisms of cytochrome oxidase activity assume considerable importance. The model presented here provides a means of studying the possible control parameters of cytochrome oxidase activity in the nervous system.

(Supported by grants from the National Institutes of Health and The Epilepsy Foundation).

HYDEN, H. *et al*. (1961). *J. Biophys. Biochem. Cytol*., *10*, 409.

TRIPHOSPHOINOSITIDE PHOSPHOHYDROLASES OF RAT BRAIN

George Hauser, Joseph Eichberg, Steven M. Gompertz
and Michael Ross

Research Laboratory, McLean Hospital, Belmont, Mass., and
Department of Biological Chemistry, Harvard Medical School,
Boston, Mass.

Brain triphosphoinositide can be degraded by a phosphomonoesterase (I) and a phosphodiesterase (II). In connection with studies on myelination and the biosynthesis of polyphosphoinositides, the aims of the present investigation were: complete separation of the two native enzymes; definition of conditions specifically inhibitory for I and II, but without effect on the biosynthesis; evaluation of possible I and II activity in purified myelin; and estimation of these activities in brain extracts throughout development.

Dialysed extracts of brain acetone powders were used for the enzyme assays. Separation of I and II was attempted by examining their extractability with water from such powders, when the length of extraction was varied. The amount of total activity (I + II in the ratio 1:1 to 1:2) was unchanged between 3 min and 3 hrs, but was decreased after 6 hrs or more, principally because of loss of I. When water was replaced by buffers extraction for 3 hrs gave similar results at physiological pH, but between pH 5 and 6 very little I activity appeared. After repeated extractions at pH 5.5, additional II could be extracted at pH 7.5. An extract enriched in I could not be obtained under these conditions.

When assayed, the extracts released a maximum of about 50% of the substrate - P in 30 minutes. Freezing lowered I activity. Mg^{++} at an optimal concentration of 1 mM activated I and inhibited II. KCl (75-150 mM) increased both activities 30 to 50%. Ouabain (0.1 to 1.0 mM) had at most a slight effect, but suramin decreased P liberation by both enzymes markedly. EDTA, 1 mM, (pH 7.2) reduced I activity by 48% and II activity by 80%, while cyclohexanediamine tetraacetate inhibited both I and II 78%. This inhibition, in contrast to that by EDTA, was not antagonized by the detergent cetyltrimethyl ammonium bromide, which, by itself, stimulated I 100% and II 50%. So far it has not been possible to utilize these findings to increase the accumulation of labelled polyphosphoinositides in a cell-free system.

Virtually no activity of I or II could be detected in purified, lyophilized myelin preparations. When brain homogenates were centrifuged at 10^5 x g, both the resulting sediments and supernatant fluids contained substantial I and II activity. No appreciable difference in enzymic activity could be shown in acetone powder extracts at any age from birth on, although the levels of polyphosphoinositides increase about six-fold during brain development.

ISOLATION AND CHARACTERIZATION OF SOLUBLE BRAIN LIPOPROTEINS

Norbert Herschkowitz, Eric M. Shooter and Guy M. McKhann

Departments of Pediatrics and Genetics,
Stanford University School of Medicine, Palo Alto, Cal.

Cholesterol, glycolipids and phospholipids appear in the 105,000 x g supernatant of rat brain homogenates in higher concentration than one would predict on the basis of the solubility of these lipids in water. The studies in the present report indicate that these lipids are bound to proteins to form water-soluble lipoprotein complexes. These soluble brain lipoproteins are distinct from those of serum in their density, as determined by preparative ultracentrifugation, their lipid composition, and in their pattern of migration on acrylamide gel electrophoresis. In addition, antibodies to both high and low density serum lipoproteins do not cross react with soluble brain lipoproteins.

In order to determine if soluble brain lipoproteins were the result of the random combination of lipid with protein during homogenization, or were a manifestation of particulate contamination, 17 day old animals were injected with ^{35}S-sulphate, and the incorporation of labelled sulphate into the sulphatide of the homogenate and the different subcellular fractions determined at 2, 19, 48, 72 and 96 hours. The temporal pattern of labelling of the sulphatide of soluble brain lipoproteins differed from those of both the whole homogenate and the myelin and microsomal fractions.

The isolation of soluble brain lipoproteins, containing ^{35}S-labelled sulphatide was accomplished by dialysis and separation by preparative acrylamide gel electrophoresis. By this technique, distinctive lipoprotein complexes, giving single bands on subsequent analytical acrylamide gel electrophoresis, could be separated. The amino acid composition and lipid composition of these isolated lipoproteins is under investigation.

These studies indicate that soluble brain lipoproteins exist as distinct entities, are not artifacts of the homogenization procedure or particulate contamination, and can be separated and characterized in terms of protein and lipid composition.

A POTASSIUM-INDUCED STIMULATION OF POTASSIUM AND AND GLUTAMATE TRANSPORT IN BRAIN SLICES

L. Hertz and T. Arnfred

Department of Biochemistry A, University of Copenhagen.

It has been suggested repeatedly that the increased rate of oxygen uptake by brain slices in a potassium-rich medium could result from cation movements across cell membranes (e.g. HERTZ, 1965). It was therefore investigated by means of curves showing the uptake and 'desaturation' (WINEGRAD & SHANES, 1962) whether the efflux and influx rates of ^{22}Na and ^{42}K were affected when high concentrations of potassium were added to a 'physiological' suspending medium. The sodium fluxes were found to be essentially unchanged, whereas both potassium in- and efflux were increased more than corresponded to the passive increase in diffusion. That this was probably due to an active process is supported by the finding that the stimulation of ion transport was inhibited by dinitrophenol or by anoxia.

Potassium ions must either be exchanged against other cations, or they must be accompanied by anions (e.g. chloride or glutamate). No stimulation of the chloride efflux occurred, however, after increasing the potassium concentration, but the efflux of glutamate- or its metabolites - was enhanced; such a coupling between the transport of glutamate ions and of potassium ions in brain slices was previously observed by TERNER, EGGLESTON & KREBS (1950).

The increased disappearance of radioactive glutamate from the tissue when it was exposed to high concentrations of potassium cannot represent simply (or only) an increased oxidation of glutamate to CO_2, since the amount of radioactivity represents unchanged glutamate or some of its derivatives is being studied chromatographically.

HERTZ, L. (1965). Nature, Lond., 206, 1091. TERNER, C., EGGLESTON, L.V. & KREBS, H.A., (1950). Biochem. J., 47, 139. WINEGRAD, S. & SHANES, A.M. (1962). J. gen. Physiol, 45, 371

GANGLIOSIDES IN RELATION TO HISTOLOGICAL STRUCTURE OF CEREBRAL CORTEX AND OF PHOTORECEPTOR ORGANELLES*

Helen H. Hess, Norman H. Bass and Charles N. Still**

The McLean Hospital Research Laboratories, Belmont and the Department of Neurology, Harvard Medical School, Boston, Mass.

The intralaminar distribution of ganglioside N-acetylneuraminic acid (G-NANA) was determined in human prefrontal and rat somatosensory cortex by serial frozen section sampling and microchemical assay. Lipids were extracted and partitioned (HESS & THALHEIMER, 1965) to obtain upper phase solids for analysis of NANA by the fluorometric method of HESS and ROLDE (1964). In man, G-NANA of cortical gray matter and immediately subjacent white matter averaged 8.3 and 3.9 μg/mg dry wt., respectively; values in rat brain were similar. The mean for subcortical white matter was appreciably above that for bovine corona radiata (0.8). The intracortical pattern in both types of cortex showed three main peaks or inflections at the junctions of the following layers: II-IIIa, IIIc-IV, and V-VIa. Layer I was the site of the highest values for NANA per unit dry wt. in human but not rat cortex. However, in rat cortex DNA analyses and histological enumeration of relative numbers of neurons and total cells demonstrated that the G-NANA/neuron cell body present was by far the greatest in layer I (23 μμg) and least in Va and the middle of VI (6.5 and 8.5). The high NANA/neuron in layer I reflects the fact that dendrites from all deeper layers terminate here in relation to axons of extra- and intra-cortical origin, while neuron cell bodies are scarce.

Frog retinal rod outer segments (ROS) were detached by spinning retinas in saline and partially purified by low speed centrifugation. Analysis of upper phase lipids by the thiobarbituric acid method (WARREN, 1959) with a dichromatic correction for large amounts of an interfering substance revealed no G-NANA. Treatment of whole ROS for 1 hr at 80° with 0.1 N-H_2SO_4 gave a total sialic acid value of 0.9 μg/mg protein. Most if not all the sialic acid may therefore exist as a polysaccharide-protein complex, not in photoreceptor disk membranes but in the enclosing membrane that comprises only 1% of the outer segment mass.

It is concluded that the distribution of gangliosides in cortex is in accord with a localization in highly arborized dendritic and axonal plexuses and their synaptic articulations – i.e., in neuronal surface membranes. Their absence from photoreceptor disks, which are derivatives of the plasma membrane of a neuron (the rod cell) may indicate exclusion of gangliosides from compound membranes, even those of neuronal origin.

HESS, H. & ROLDE (1964). J. biol. Chem., 239, 3215. HESS, H. & THALHEIMER (1965). J. Neurochem., 12, 193. WARREN (1959). J. biol. Chem., 234, 1971.

*Supported by USPHS Grant No. NB-00361; NSF Grants GB-574 and GB-3128; and by a grant from the Scottish Rite Committee on Research on Schizophrenia of the National Association for Mental Health.

**Present address: William S. Hall Psychiatric Institute, Columbia, S.C.

NEUROCHEMICAL EFFECT OF NORMAL ENVIRONMENTAL DEPRIVATION

Williamina A. Himwich and Harish C. Agrawal

Thudichum Psychiatric Research Laboratory,
Galesburg State Research Hospital, Galesburg, Illinois

Experiments with animals reared in different environmental conditions have shown that social contact with other animals or mother during maturation is needed for the development of a normal pattern of behaviour and biochemical maturation of the brain. Moreover, a few studies have been made on the effect of handling, visual deprivation and stimulation on the enzymes and protein contents of the nervous tissue and most of the workers favour the view that a certain amount of physiological stimulation is necessary for the normal development of nervous tissues, and that alterations in environmental condition and deprivation interfered with the metabolism of amino acids and proteins. We have, therefore, investigated the effects of isolation, maternal deprivation and light deprivation on amino acids and biogenic amines of some discrete areas of dog, rat and rabbit brain respectively as described in the following studies: (1) Young puppies reared in partial isolation for one week from 4-5 weeks of age showed behavioural abnoramlities, being hyperactive with diffuse reactions. Significant changes also occurred in the concentration of free amino acids, chiefly of glutamic acid origin, in the subcortical areas. (2) Similarly, maternal deprivation of rats at 15 and 19 days of age had a pronounced effect on amino acids and acetylcholinesterase activity of suckling rats. Glutamic acid derivatives as well as alanine and threonine exhibited considerable increase in cerebral cortex and cerebellum while a more or less proportionate decrease in the level of these amino acids was observed in mid-brain.

REGIONAL DIFFERENCES OF RETINAL LDH-PATTERNS AND THEIR FUNCTIONAL IMPLICATION

C.G. Honegger

Neurological Clinic, University of Basel, Switzerland

A study of the LDH-isoenzyme patterns in the retinal layers of different mammals was carried out. The densitometric evaluation revealed in the avascular retina (rabbit and horse) a characteristic asymmetric isoenzyme pattern. Relatively high H/M-values were observed especially in the region - outer nuclear layer to inner segments - with a high proportion of LDH_1. This was not seen in the vascularized retina of cow and monkey. Our knowledge of the isoenzyme distribution in other tissues suggests that these regions contain cell types with different isoenzyme patterns. According to histochemical observations the LDH-activity is localized mainly in the radial glia and it shows a higher activity in the avascular retinas. These results indicate metabolic differences of the radial glia in avascular and vascularised retinas. In addition, differences in the LDH-isoenzyme patterns from distinct regions of cow retinas were found. From these results, as well as from those of the retinas of iodacetate poisoned rabbits, the LDH-isoenzyme distribution in relation to the specific function of this tissue will be discussed.

THE SHORT-TERM TURNOVER OF ETHANOLAMINE PHOSPHOGLYCERIDES IN MOUSE BRAIN MYELIN AND MICROSOMES IN VIVO

Lloyd A. Horrocks

Laboratory of Neurochemistry,
Cleveland Psychiatric Institute, Cleveland, Ohio.

Brain myelin in developing animals incorporates a considerable amount of a labelled precursor. Intracerebral injection of [^{14}C]ethanolamine provides a method of incorporating large amounts of radioactivity into brain ethanolamine lipids (HORROCKS & ANSELL, 1967). The aim of the present study was to determine the metabolic relationships of ethanolamine phosphoglycerides in mouse brain myelin and microsomes.

Forty day old mice (C57BL/10 strain) taken from 3 litters were injected intracerebrally with 10 µl of a solution containing 0.25 µcuries of [^{14}C]ethanolamine. Two mice from each litter were sacrificed at 1, 4, and 7 days after injection. Lipid extracts from myelin and microsomal fractions were prepared and characterized. The acid-labile EPG (ethanolamine plasmalogens) were characterized separately from the other ethanolamine phosphoglycerides.

An analysis of variance was made of the specific radioactivities of myelin acid-stable EPG, myelin acid-labile EPG, microsomal acid-stable EPG, and microsomal acid-labile EPG. Significant time ($F = 25.8$, df 2/15, $P<0.0001$), type of compound ($F = 24.1$, df 3/45, $P<0.0001$) and interaction ($F = 4.59$, df 6/45, $P<0.005$) differences were found. It was shown that each of the 4 types of compounds had significantly lower specific radioactivities at day 7 when compared with either day 1 or day 4 by multiple range tests ($P<0.01$). For example, myelin acid-labile EPG had specific radioactivities of 5700, 9300, and 2500 dpm/µmole at days 1, 4, and 7. Myelin/microsome ratios of specific radioactivities were also subjected to an analysis of variance. The ratios increased with time ($F = 6.35$, df 2/15, $P<0.025$) showing that the microsomal EPG turned over at a faster rate than did the myelin EPG. MANDEL and NUSSBAUM (1966) also found a gradual increase of the myelin/microsomal EPG ratios with time using ^{32}P as a precursor. It was concluded that the myelin ethanolamine phosphoglycerides incorporate and turnover radioactivity at a considerable although lower rate than the microsomal ethanolamine phosphoglycerides.

Supported in part by Grant NB-05510, U.S. Public Health Service.

HORROCKS, L.A. & ANSELL, G.B. (1967). Submitted for publication. MANDEL, P. & NUSSBAUM, J.L. (1966). J. Neurochem., **13**, 629.

THE PHOSPHOLIPID COMPOSITION OF CELL FRACTIONS FROM THE MUSCLE OF NORMAL AND DYSTROPHIC MICE

B.P. Hughes and K. Owens

Department of Chemical Pathology, Institute of Neurology, The National Hospital, Queen Square, London, W.C.1.

It has been shown previously (HUGHES, 1965) that muscle from mice with a hereditary myopathy (Bar Harbor strain) contains relatively less lecithin and more sphingomyelin than muscle from normal controls. The difference is small but statistically significant. To ascertain whether these findings result from a uniform change in the composition of phospholipid-rich cell structures, or represent changes in a particular structure or group of structures, the phospholipid composition of the following fractions was examined.

Muscle was homogenised in a 0.25 M-sucrose medium essentially as described by PENNINGTON (1961) and three fractions prepared by centrifugation: (a) 600 ppt, a myofibrillar fraction, (b) 12000 g ppt, a mitochondrial fraction, (c) 35000 g ppt. Fraction (c) presumably contains elements of the sarcoplasmic reticulum but in addition estimation of succinate-INT reductase activity revealed contamination with mitochondria or mitochondrial fragments, particularly in the case of dystrophic tissue. However, this could be considerably reduced by using shorter homogenisation times. Phospholipid analysis of extracts prepared in chloroform-methanol, 2:1, by one and two dimensional thin layer chromatography on silica gel (HUGHES & FRAIS, 1965; OWENS, 1966) revealed no significant differences between normal and dystrophic muscle with respect to fractions (a) and (b). In the case of (c) however, fractions prepared from dystrophic muscle contained a smaller proportion of lecithin and more sphingomyelin compared with muscle from unaffected animals.

HUGHES, B.P. (1965). Research in Muscular Dystrophy. Proceedings of the 3rd Symposium, p. 187, Lond: Pitman Medical.
HUGHES, B.P. & FRAIS, F.F. (1965). Biochem. J., 96, 6P.
OWENS, K. (1966). Biochem. J., 100, 354. PENNINGTON, R.J. (1961). Biochem. J., 80, 649.

TURNOVER RATES OF PROTEIN AND RNA IN SUBCELLULAR FRACTIONS OF RAT BRAIN

Kern von Hungen, Henry R. Mahler and Walter J. Moore

Chemical Laboratory, Indiana University, Bloomington

The time course of labelling in carefully separated subcellular fractions from rat cerebral cortex following intraventricular injection of [^{14}C]orotate and [^{3}H]leucine was followed from 4 hr to 70 days. Subcellular fractionations were made by sucrose-gradient centrifugation and the fractions controlled by electron-microscopy and enzymic assays. The synthesis and breakdown of protein and RNA may follow complex pathways, but the reactions of labelling and delabelling could be represented as first-order processes with constants k_1 and k_{-1}, combined with a diffusive loss of label from the cell with transfer coefficient k_D. After initial transients the decline in label will be exponential with an apparent rate constant $k = k_{-1} k_D / (k_1 + k_D + k_{-1})$. If k_D is considerably greater than $k_1 + k_{-1}$, the effective k reduces to k_{-1}, the specific rate of breakdown of the labelled protein or RNA, where the half-life $\tau = \ln 2 / k_{-1}$. In various fractions (nuclei excepted) the declines in activity after 10 days followed approximately exponential curves with single time constants.

The τ for the RNA fractions was the same for all fractions, within experimental reproducibility, viz., $\tau = 12 \pm 1$ d. The ratio of ^{14}C / ^{3}H activities in the ribosomal fractions was a constant during the decline in activity. The results for the τ(protein) of various fractions were as follows: Mitochondria 19 d.; synaptosomal mitochondria 21 d.; nerve end particles (synaptosomes) 19 d.; ribosomes 12.5 d. The microsomal fraction showed definite dispersion of τ's, with an 8 d. and a >14 d. component. In the synaptosome fraction there was a delay in uptake of label during the first few days. This behaviour supports the concept of transport of macromolecules from soma to synaptic regions in neurons.

The $\tau \sim 20$ d. for mitochondrial proteins in brain compares with $\tau \sim 10$ d. (FLETCHER & SANADI, 1961) for rat-liver mitochondria. Comparisons between other brain and liver fractions confirm this generally lower rate of protein turnover in the brain. Nevertheless, *in vitro* preparations of rat-brain polysomes and mitochondria both show higher optimal protein incorporating activities than the optimal activities of corresponding liver fractions. This comparison suggests that brain cells may have a reserve capacity for rapid protein synthesis.

One of the aims of this study was to determine the half life of synaptic-vesicle protein in order to secure evidence as to possible precursors of these entities. The half life $\tau = 19$ d. for the vesicles is about the same as that of other subcellular fractions. This long half life tends to support a model in which vesicles are used repeatedly during the firing of synapses, rather than being destroyed at each firing.

These data will be compared with our data for sub-cortical and cerebellar regions of rat brain.

FLETCHER & SANADI (1961). *Biochim. Biophys. Acta*, 51, 356.

THE BIOSYNTHESIS OF RNA AND PROTEIN IN NORMAL AND SCRAPIE MOUSE BRAIN

G.D. Hunter and R.H. Kimberlin

A.R.C. Institute for Research on Animal Diseases, Compton, Newbury, Berks., England.

Studies carried out *in vivo* have shown that there is little difference between the overall rates of protein and RNA synthesis in normal and scrapie mouse brain. In both cases, the intracerebral route of inoculation of labelled intermediates is usually to be preferred to the intraperitoneal route. Certain lysosomal enzymes, notably β-glucuronidase and N-acetyl-β-D-glucosaminidase do, however, appear to be synthesised at elevated rates in scrapie mouse brain (HUNTER & MILLSON, 1966).

More detailed analysis of RNA synthesis in mouse brain has so far largely been confined to material obtained from normal mouse brain. At early times after the injection of $^{32}PO_4^{3-}$ or labelled uridine, the rapid synthesis of RNA can be detected in brain and much of this RNA is more DNA-like in base composition than the ribosomal RNA. At later times, it is only possible to detect the synthesis of ribosomal RNA, which appears to turnover at the rate of about 5% per day (KIMBERLIN, 1966).

In vitro, preparations of ribonucleoprotein particles from mouse brain incubated under suitable conditions incorporate amino acids into protein at rates comparable to those observed using similar preparations of ribonucleoprotein particles from liver (HERRIMAN & HUNTER, 1965). The stimulation of the activity of these particles by homologous brain RNA has been studied in some detail, mainly using preparations derived from normal brain. In general the messenger theory seems to be applicable to the situation observed in mouse brain. Messenger RNA activity has been found in fractions eluted from methylated albumin columns, and also in RNA fractions isolated from sucrose density gradient runs. In the latter case, the most active fraction was that which showed the highest rate of labelling at early times after the intracerebral inoculation of isotopic precursors of RNA.

HERRIMAN, I.D. & HUNTER, G.D. (1965). *J. Neurochem.*, *12*, 937.
HUNTER, G.D. & MILLSON, G.C. (1966). *J. Neurochem.*, *13*, 375.
KIMBERLIN, R.H. (1967). *J. Neurochem.* In the press.

COMPARATIVE STUDIES OF THE DEAMINATION OF DIMETHOXYPHENYLETHYLAMINE IN RAT, DOG AND HUMAN BRAIN

Z. Huszti and J. Borsy

Department of Pharmacology,
Research Institute for Pharmaceutical Chemistry, Budapest.

Dimethoxyphenylethylamine (DMPE) has aroused considerable interest since its identification in the urine of acute schizophrenic patients (FRIEDHOFF *et al.*, 1962). Presumably it is one of the products of the defective metabolism of dopamine (FRIEDHOFF *et al.*, 1963). The question whether this compound is characteristic of schizophrenics, or whether it appears in normal individuals as well, is still open (TAKESEDA *et al.*, 1963; SENN *et al.*, 1964). However there is suggestive evidence that the metabolism of biogenic amines in schizophrenics differs from that in normal individuals. In view of this the rate of degradation of DMPE in brain homogenates was studied in detail.

The first studies carried out *in vitro* on rabbit and rat liver (HUSZTI, 1965; HUSZTI & BORSY, 1966) and the *in vivo* studies in rats and humans (SCHWEITZER *et al.*, 1966; FRIEDHOFF *et al.*, 1966) suggested that the degradation of DMPE is due mainly to the amine oxidase.

Recent experiments with rat, dog and human brains have shown that the deamination of DMPE is effected also by brain tissue, but the rate of deamination depends on the species or, with human brain, on the physiological state. The respective specific activities of rat, dog and human brain homogenates, using DMPE as substrate, were 0.50, 0.03 and 0.30 μatoms oxygen/mg protein/60 min. The deamination of DMPE is effected mainly in the first and second fractions obtained by fractionating brain cells according to LØVTRUP (1961). No deamination of DMPE was observed in the supernatant fraction, containing microsomes and cytoplasm. A comparison of the rate of deamination of DMPE and serotonin by different cell fractions showed no parallelism between the two substrates. Using different cell fractions the serotonin/DMPE ratio varied slightly. However the degradation of DMPE was inhibited by the well-known inhibitors of the monoamine oxidase.

The differences between the degradation of DMPE in schizophrenics and normal individuals will be discussed.

FRIEDHOFF, A.J. & VAN WINKLE, E. (1962). *Nature, Lond.*, 194, 897. FRIEDHOFF, A.J. & VAN WINKLE, E. (1963). *Nature, Lond.*, 199, 1272. FRIEDHOFF, A.J. & HOLLISTER, L.E. (1966). *Biochem. Pharmacol.*, 15, 269. HUSZTI, Z. (1965). *Int. Neurochem. Conference, Oxford*, 1965. HUSZTI, Z. & BORSY, J. (1966). *Biochem. Pharmacol.*, 15, 475. LØVTRUP, S. & ZELANDER, T. (1962). *Experimental Cell Res.*, 27, 468. SCHWEITZER, J.W. & FRIEDHOFF, A.J. (1966). *Biochem. Pharmacol.*, 15, 2097. SENN, N.P. & MCGEER, P.L. (1964). *Biochem. Biophys. Res. Comm.* 14, 227. TAKESEDA, M.H., KAKINOTA, Y., SANO, Y. & KANETO, Z. (1963). *Nature, Lond.*, 199, 203.

SIMULTANEOUS DETERMINATION OF RNA AND PROTEIN SYNTHESIS IN RAT BRAIN CORTEX SLICES

Matti Huttunen and Johan Järnefelt

Department of Medical Chemistry,
University of Helsinki, Finland.

Experiments are described in which the simultaneous incorporation of [^{14}C]orotic acid into RNA and of [^{3}H]leucine into protein in brain slices was studied. These studies were initiated in order to find conditions for the study of the effects of various functional states on the protein and nucleic acid metabolism of nervous tissues.

Rat brain cortex slices were cut in the cold, weighed and transferred into glycylglycine-buffer containing 10 mM-glucose, 0.1 mM-leucine and orotic acid and an amino acid mixture corresponding to that of cerebrospinal fluid. After 15 min preincubation 2.5 µC [^{14}C]orotic acid and 10 µC [^{3}H]leucine were added. The incubation was carried for 45 min, during which oxygen consumption was measured. After homogenization and washing with 5% TCA, alcohol, alcohol-chloroform, and ether, the nucleic acids were hydrolysed with hot 5% TCA, and the RNA, DNA and ^{14}C-content in the supernatant were determined. The remaining precipitate was washed and dried. The protein and ^{3}H-content were determined (MAHIN et al., 1966). In control experiments there was practically no incorporation of ^{3}H into RNA or ^{14}C into protein.

In slices cut serially from the cortex, the superficial slice always showed about 1.5 times greater incorporation of both orotic acid and leucine than the next slice. The incorporation of both leucine and orotic acid was clearly stimulated by the presence of an amino acid mixture. During incubation the pH of the medium dropped about 0.2 - 0.3 pH units. It was found that an initial pH of 7.7 in the medium produced optimal incorporation, which was 2 - 3 times higher than with an initial pH of 7.4.

MAHIN et al. (1966). Anal. Biochem., 16, 500.

UTILIZATION OF ADENOSINE BY CALF BRAIN EXTRACTS

P.L. Ipata

"Impresa Enzimologia" of the C.N.R.
Institute of Biological Chemistry, Università di Pisa,
Pisa, Italy

It has been shown by MANDEL (1966) that AMP does not penetrate the brain barrier, but undergoes a prior dephosphorylation to adenosine; the nucleoside is readily incorporated into RNA of the central nervous system without cleavage of the purine ring.

This communication describes the utilization of adenosine by calf brain extracts. Adenosine is converted to IMP, the common precursor for ATP and GTP synthesis, by the following pathway:

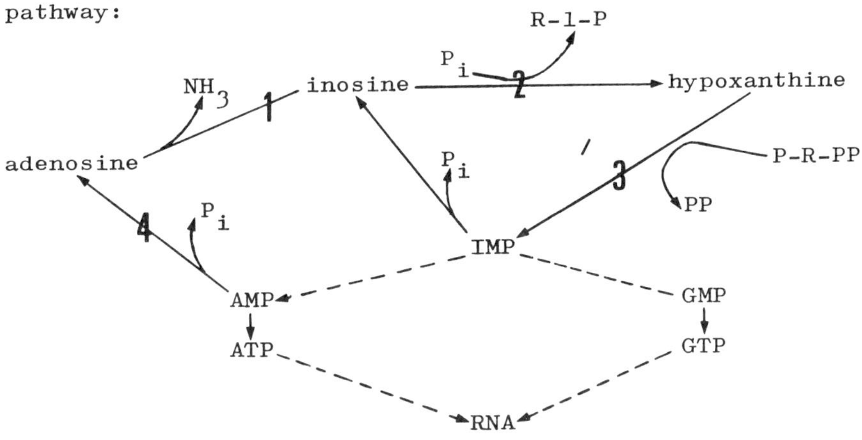

1 = adenosine deaminase; 2 = nucleoside phosphorylase;
3 = IMP pyrophosphorylase; 4 = 5-nucleotidase.

Enzyme 4 was extensively purified, and shown to be strongly inhibited by ATP and other nucleoside-triphosphates, except GTP. On the base of the following observations it can be concluded that the inhibition is of the "allosteric" type: (a) sigmoid form of inhibition curve obtained when the enzyme was assayed in the presence of increasing concentrations of nucleoside-triphosphates; (b) the value of the interaction coefficient, n', for each nucleoside-triphosphate was significantly greater than 1; (c) 'desensitization' to ATP inhibition was obtained by treatment with pCMB; (d) mixing two nucleoside-triphosphates resulted in inhibition significantly exceeding the sum of their fractional inhibitions. The significance of 5-nucleotidase inhibition by nucleoside-triphosphates will be discussed.

In the above scheme it can be seen that adenosine is converted back to adenosine via enzymes 1, 2, 3 and 4, and inosine back to inosine via enzymes 2, 3 and 4. The inhibition by ATP and other nucleoside-triphosphates on the 5-nucleotidase (acting on both AMP and IMP) would therefore avoid the occurrence of the two 'futile metabolic cycles' and the wasteful breakdown of ATP needed for P-R-PP synthesis. The situation is similar to that reported by GEVERS and KREBS (1966) for F-1-6-Pase inhibition by AMP.

GEVERS & KREBS (1966). Biochem. J., **98**, 720. MANDEL (1966). Proc. Brain Barrier Conference, Amsterdam, 26 - 29 Sept.

PREPARATION OF BULK QUANTITIES OF DIFFERENTIATING CELLS FROM THE CHICK EMBRYO CEREBRAL HEMISPHERE

M. Jacob, C. Judes and M. Sensenbrenner

Centre de Neurochimie du CNRS and Institut de Chimie Biologique, Faculté de Médecine, Strasbourg

In order to perform the biochemical experiments necessary for a better understanding of differentiation, adequate or sufficient quantities of cells of the same type must be available. A technique of bulk separation of differentiating cells of the chick embryo hemisphere is presented. The tissue was treated by a mixture of glycerol-acetone-water (1:1:1 by volume) according to SATAKE and ABE (1966). The cells were dissociated by passage through a nylon sieve of 82 μ. They were identified by comparison with histological sections of hemispheres of embryos of the same age. The proportion of the various cell types at several ages were also determined.

In hemispheres of 6 day old embryos, about 10 per cent of cells were young unipolar neuroblasts and 90 per cent were undifferentiated cells with a big nucleus and very little cytoplasm. At 12-13 days of age, 34 per cent of the cells were neuroblasts and amongst them many were multipolar. The other cells were morphologically close to the undifferentiated cells. After 13 days of age, no further increase of the relative proportion of the neuroblasts and small cells could be detected. The neuroblasts increased in size and carried more processes while the small undifferentiated cells became differentiated into spongioblasts and glial cells.

By centrifugation in discontinuous sucrose gradients (1.0 M, 1.3 M, 1.9 M) it was possible to separate the neuroblasts from the spongioblasts (or small cells) of the hemispheres of 12-13 day old embryos. These fractions contained 80 per cent of neuroblasts or 90 per cent of spongioblasts and small cells. In older embryos the same type of fractionation could be applied with similar results.

A study of the biosynthesis of nucleic acids during differentiation of neural cells has been undertaken, starting with the indifferent cells of the 6 day old embryos.

TRANSPORT OF MONOSACCHARIDES INTO SLICES OF GUINEA-PIG BRAIN CORTEX

P. Joanny, J. Corriol, A. Kleinzeller and H.H. Hillman

Laboratoire de Physiologie, Faculté de Médecine, Marseille

Guinea-pig brain cortex slices were incubated for 35 min at 37°C in Krebs-Ringer bicarbonate saline in the presence of 20 mM-lactate as substrate and various concentrations of the sugars to be tested. After correction for the extracellular (inulin) space, i.e. 50% of tissue wet weight, practically no free glucose was detected in the apparent intracellular fluid under either aerobic or anaerobic conditions at the glucose concentrations tested (1 to 20 mM). Aerobically, in the presence of 1 mM-iodoacetamide, glucose was found in the total intracellular fluid. Thus, the rate of glucose entry into the cells appears to be much slower than its intracellular utilization. With various concentrations of such non-metabolizable sugars as 2-deoxy-D-glucose and radioactive D-xylose, S_I/S_O ratios were close to 1.0. On pre-incubation of slices for 20 min in ice-cold saline, the sugars were confined to the extracellular space; subsequent aerobic incubation at 37° allowed the determination of the kinetics of sugar entry into the cells. The initial rate of transport of most sugars showed saturation kinetics. The sugars D-glucose (1 mM-iodoacetamide) with a K_m of 125 mM and a V_{max} of 20 mM/l cell H_2O/min, and 2-deoxy-D-glucose with a K_m of 57 mM and a V_{max} of 9 mM/l cell H_2O/min showed the highest apparent affinities. Much less affinity was shown by D-xylose with a K_m of 222 mM and a V_{max} of 29 mM/l cell H_2O/min, and D-arabinose, L-sorbose and D-fructose. The values with D+mannose, D-galactose and D-ribose plotted on a Line-weaver-Burk plot gave a straight line going through the origin suggesting diffusion kinetics. A competition for their entry into the cells was demonstrated between D-glucose and 2-deoxy-D-glucose, and between D-xylose and D-glucose. Most of the monosaccharides thus stem to penetrate the cells as free sugars by facilitated diffusion, no accumulation against a concentration gradient taking place.

BRAIN PROTEINS LABELLED BY [^{32}P]DFP AND THEIR CONNECTION WITH THE DELAYED NEUROTOXIC EFFECT OF SOME ORGANOPHOSPHORUS COMPOUNDS

M.K. Johnson

Medical Research Council Toxicology Research Unit, Carshalton, Surrey.

Tri-ortho-cresyl phosphate (T.O.C.P.), di-isopropyl phosphofluoridate (DFP) and numerous organophosphorus compounds produce ataxia of hind limbs in many species including man, and a characteristic feature is the delay of 7-14 days after administration of the compound before any effect is seen. Attempts to correlate neurotoxic activity with chemical structure or with the inhibition of known enzymes *in vivo* have failed (see ALDRIDGE & BARNES, 1966). Organophosphorus compounds phosphorylate some proteins forming a stable covalent bond. It is proposed that [^{32}P]DFP will in this fashion label some proteins of the nervous system when it is administered in a neurotoxic dose to hens and that a protein (?esterase) involved in the neurotoxic effect will be among those labelled.

Ninety three per cent of those proteins of whole hen brain which are labelled with [^{32}P]DFP *in vivo* (1 mg/kg) or *in vitro* (10^{-5} M for 10 min at 25°/pH 8.0) can be brought into solution by treatment of a homogenate with sodium deoxycholate (0.8% final in Tris chloride pH 8.0). Proteins are separated from unchanged DFP by passage through a Sephadex G-25 column loaded with a neutral detergent (B.D.H. Brij 35) in the same buffer. The effluent is concentrated and labelled proteins separated by chromatography on DEAE cellulose at pH 8.0 using NaCl for gradient elution. Buffer and detergent are chosen so that elution of protein can be monitored by absorption at 225 mμ (WADDELL, 1956). Radioactivity is measured with 37% efficiency and without loss of specimen by scintillation in water instead of standard medium. Results of separations by chromatography and electrophoresis will be presented. Progress towards pinpointing the significant protein has been simplified by preincubation of brain homogenates with non-neurotoxic organophosphorus compounds which are active inhibitors of many enzymes and by administration of such compounds to the hen *in vivo* before treatment of homogenates with [^{32}P]DFP *in vitro*.

ALDRIDGE, W.N. & BARNES, J.M. (1966). Biochem. Pharmacol. 15, 541 & 549. WADDELL, W.J. (1956). J. Lab. Clin. Med., 48, 311.

LIPIDS AND FREE AMINO ACIDS IN HUMAN BRAIN SLICES

D. Karcher, K. Adriaenssens and A. Lowenthal

Department of Neurochemistry, Born-Bunge Foundation, Berchem-Antwerp

The heterogeneity of the central nervous system impairs biochemical examinations of the human brain. The need to study nervous tissue sections taking the anatomical areas into consideration has often arisen.

In view of this, techniques have been developed for examining the lipid composition by thin layer chromatography and the amino acids by paper chromatography using tissue sections varying between 20 and 240 mμ in thickness. These methods can be used to identify phospholipids, glycolipids, cholesterol and cholesterol esters and also about 20 amino acids. They can be applied to brain biopsy material as well as to any part of the central nervous system. A histological control can be carried out after the biochemical examination.

BINDING OF METALLIC CATIONS BY AQUEOUS DISPERSIONS OF BRAIN LIPIDS

R. Katzman, M.B. Abramson, R. Curci and H.P. Gregor

The Saul R. Korey Department of Neurology,
Albert Einstein College of Medicine, Bronx, New York

In brain tissue, acidic lipids constitute a significant proportion of the anionic sites, as shown by FOLCH-PI and others. We have studied the reactions of metallic cations with highly purified phospho- and glycolipids dispersed in water by ultrasonic radiation. The role of these lipids in determining the distribution of metallic cations must depend both upon the compartment in which they are located (plasma membrane, cytoplasm, mitochondria, nucleus, etc.) and upon their specific binding properties relative to that of other tissue anions. The binding of cations to the lipids at neutral pH was measured by use of titration, hydrogen ion release, light scattering; and stability or formation constants (K_f) of the lipid-cation complexes were measured. The acidic phospholipids, phosphatidic acid (PA), phosphatidylserine (PS), and phosphatidyl inositol (PI) bind divalent cations (Ca>Mg) strongly; more than most soluble tissue phosphates. Thus, the K_f of CaPA (1.6×10^4) is greater than that of Ca to ATP and very much greater than that of Ca to AMP, ADP, or PCr. In contrast, the binding of Mg to PA (1.2×10^4) is less than that of Mg ATP. PS and PI are less strongly bound to Ca than PA and are more like ATP, but again with the Ca over Mg preference. Sulphatide-like sulphonic acid resins show much less binding than the phospholipids, but still show the order of Ca>Mg. The binding of univalent cations by the phospholipids is much less than for the divalent. Here, the order is Li>Na>K>TMA. The K_f of NaPA is 9, 1/1000 that of the Ca. The relative selectivity of Na/K is about 2:1 which is greater than that of most analogous ion exchange resins. Sulphatide shows the reverse, K being preferred over Na.

Can this data be extrapolated to the prediction of lipid-cation distribtuion in a compartment containing a mixture of cations, lipids and soluble organic anions? In dispersions containing a single lipid and a mixture of cations, the predicted binding holds up well and has been confirmed by direct analysis of such mixtures. The presence of soluble organic anions with strong binding properties, e.g., citrate, does not alter the K_f of the lipid-cation. However, when mixtures of lipids are studied, pronounced changes are found. Thus, lecithin reduces the binding of divalent cations to monovalent lipids. This effect may in part be due to the fact that if lecithin is dispersed in the micelle then acidic lipid groups are so separated that a single divalent cation can no longer combine with two adjacent lipid molecules. Under such circumstances if the other valence of cation were to be neutralized by a soluble anion, the cation could serve to bring other molecules in close proximity to the lipid surface.

INTERRUPTION OF HYPOGLYCEMIC COMA BY INFUSION OF AMINO ACIDS

D. Kemali, F. Cioffi and F. Smaltino

The Nervous and Mental Clinic of the University of Naples.

Psychotic patients were investigated in a study of the correlation between clinical EEG findings and the ability of certain amino acids injected intravenously to produce a regression of insulin hypoglycemic coma. The amino acids: L-aspartic acid, L-glutamic acid, L-alanine, L-citrulline, L-ornithine, L-arginine, DL-threonine and L-phenylalanine, were injected either singly or in various combinations. In grouping the different amino acids consideration was given to their biochemical functions: transamination processes, urea synthesis and glycogenesis.

The changes in the perhipheral blood sugar level did not appear to be closely related to the state of consciousness during recovery from coma.

ELECTROCHEMICAL POTENTIAL DIFFERENCES FOR H^+/HCO_3^- BETWEEN BLOOD AND CSF IN ACUTE AND SUSTAINED ACID-BASE CHANGES

Å. Kjällquist and B.K. Siesjö

Department of Neurosurgery, University of Lund, Sweden

It has previously been shown that H^+ and HCO_3^- ions are not distributed at electrochemical equilibrium between blood and cerebrospinal fluid (CSF), and further that the electrical potential between the two phases varies linearily with plasma pH in acute experiments. There is, however, a controversy regarding the potential changes in chronic acid-base shifts. Since information on electrochemical potential differences is essential for evaluation of active transport of H^+ across the blood-CSF barriers, a series of rat experiments were made with measurements of both the ion concentrations in the two phases and the electrical potential difference between them. The acid-base parameters in blood and CSF were measured by using microelectrodes for pH and pCO_2, and a microdiffusion method for CSF bicarbonate. The electrical potential difference between blood and CSF was measured with a differential amplifier and calomel reference electrodes with KCl bridges. Blood and CSF were drawn into plastic housings to make contact with the KCl bridges. The CSF was drawn from the suboccipital cistern after puncture of the atlanto-occipital membrane with a fine glass pipette.

The experiments included acute respiratory and non-respiratory acid-base shifts as well as chronic nonrespiratory acidosis and alkalosis. By combining respiratory and non-respiratory acid-base changes it was possible to study the effect of hypercapnia on the CSF H^+, either at constant plasma bicarbonate or at constant pH (constant blood-CSF potential). On the basis of the results a discussion is given about passive and active factors influencing the concentrations of H^+ and HCO_3^- in the extracellular fluids of the central nervous system.

HISTOCHEMICAL DISTRIBUTION OF ACETYLCHOLINESTERASE IN SYMPATHETIC GANGLIA OF CONTROL AND IMMUNOSYMPATHECTOMIZED RATS

Gerda I. Klingman

Department of Biochemical Pharmacology,
School of Pharmacy, State University of New York at Buffalo,
Buffalo, New York

The administration of Nerve Growth Factor Antiserum to newborn rats reduces the cell populations of sympathetic ganglia by 75-93%. It was of interest to determine whether the surviving cells were essentially of the cholinergic sympathetic type. The histochemical method of Koelle was employed to study the distribution of cholinesterases in superior cervical, stellate, thoracic chain and abdominal sympathetic ganglia of control and immunosympathectomized rats. To differentiate between acetylcholinesterase and pseudocholinesterase activity, two specific inhibitors, ethopropazine[1] (1.58×10^{-5} M) and B.W. 284C51[2] (5×10^{-6} M) were used. All sympathetic ganglia of control rats, including the abdominal ganglia, contain substantial numbers of cells with marked acetylcholinesterase activity, e.g. in superior cervical ganglia 25-30% of the cells show marked acetylcholinesterase, 8-15% low acetylcholinesterase and the remainder moderate acetylcholinesterase activity. Only 7-10% of the cell population of superior cervical ganglia survived in immunosympathectomized rats. All three cell types, based on acetylcholinesterase activity, are present among the cells which persists after antiserum treatment. The present investigation has shown that the composition with regards to the acetylcholinesterase content of the surviving cell population of sympathetic ganglia, is similar in immunosympathectomized rats to that in control rats. Assuming that the sympathetic ganglia of the rat contain cholinergic cells, as it has been demonstrated in the cat, it may be concluded that the cholinergic sympathetic cells are not different from the adrenergic sympathetic cells in their requirement, during development, of Nerve Growth Factor.

Supported by grants K3-GM-5352 and AM-06594, United States Public Health Service.

[1] Parsidol, Warner Lambert Research Laboratories.
[2] Burroughs Wellcome & Company, Inc.

AMINO ACID METABOLISM IN RAT SUPERIOR CERVICAL GANGLION

Jack D. Klingman and William McBride

Department of Biochemistry, School of Medicine,
State University of New York at Buffalo, Buffalo, New York.

A rat superior cervical ganglion was removed under urethane anaesthesia, desheathed, positioned into bipolar platinum electrodes, placed in a glucose-salt bathing solution and aerated with O_2-CO_2 at 37°. Uniformly labelled [^{14}C]glucose or ^{14}C labelled amino acid was added to the bathing solution and after a given period of time, the ganglion was stimulated continuously at 5 cps. At the end of the experimental time, the ganglion was removed from the electrodes and homogenized to remove the water soluble components. This supernatant and the bathing solution were placed on separate Dowex columns and the amino acids eluted. The trifluoroacetyl-butyl esters of the amino acids were prepared and separated by our new gas chromatographic procedure. A portion of each amino acid peak was collected, placed in a scintillation counter and the radioactivity determined. Data will be presented as to the rates of labelling of the intracellular and extracellular pool of amino acids. The rate of entry of some amino acids and their metabolism to other amino acids will also be discussed.

Supported by U.S.P.H. grant NB-03697 and The Erie County Heart Association.

INTRINSIC PROTEIN-SYNTHESIZING MACHINERY OF THE AXON

Edward Koenig

Department of Physiology and Neurosensory Laboratory
State University of New York at Buffalo, Buffalo, New York.

Previous studies have shown that the mammalian axon contains RNA (KOENIG, 1965a) and synthesizes acetylcholinesterase (AChE) locally (KOENIG, 1965b). Results from recent experiments: (1) provide further evidence for axonal protein synthesis; (2) reveal the existence of a local DNA-dependent RNA synthesis; and (3) suggest that the rate of AChE synthesis may be regulated by a local DNA-dependent mechanism.

A method for analysing tritium-labelled protein solubilizates of isolated cellular material in the range of 0.5 - 500 mμgm was developed in this laboratory. Rabbit spinal accessory nerve roots were incubated *in vitro* with tritium labelled protein and RNA precursors. Microradiochemical analyses were performed exclusively on denuded, myelin-free axons.

Incorporation of [^{3}H]leucine into axonal protein proceeded at maximal rate for 10 hrs and fell off sharply thereafter. Incorporation, compared with that of control, was unaffected by: (1) preincubation for 18 hrs with actinomycin-D; and (2) incubation with chloramphenicol.

Incubation with [^{3}H]adenine and [^{3}H]orotic acid for 4 hrs resulted in incorporation into axonal RNA. Only part of the radioactivity could be recovered by ribonuclease (RNAse) digestion. Over 85% of the total radioactivity was associated with the RNAse-resistant fraction. Actinomycin-D significantly inhibited incorporation into the RNAse-resistant fraction.

Application of actinomycin D to a segment of hypoglossal nerve *in vivo* produced a significant increase in AChE activity in DFP-pretreated and in untreated cats 2 or more days later. The increase was limited to the region treated and was blocked by puromycin or 5-fluororotic acid. *In vitro* incubation of nerves with actinomycin-D produced no increase.

The results indicated that: (1) the axon contained an unusually stable protein-synthesizing system; (2) mitochondrial protein synthesis did not contribute measurably to overall incorporation; and (3) a local DNA-dependent RNA synthesis seemed operative. It is possible that protein synthesis in the axon may be concerned with renewal of the axolemma, and part of its dynamic aspects may be under regulation of a local DNA-dependent mechanism.

KOENIG, E. (1965a). *J. Neurochem.*, **12**, 357. KOENIG, E. (1965b). *J. Neurochem.*, **12**, 343.

INTERRELATIONS OF RNA AND PROTEIN SYNTHESIS AND NEURAL FUNCTION IN CAT SPINAL CORD. A STUDY OF THE NEUROBIOLOGICAL ACTIVITY OF ACTINOMYCIN D (AMD)[1]

Harold Koenig and Chung Lu

V.A. Research Hospital and Northwestern University Medical School, Chicago, Illinois

AMD, an inhibitor of DNA-directed synthesis of RNA, exerts potent neurobiological effects upon intrathecal administration (KOENIG _et al_., 1967a; KOENIG & JACOBSON, 1967). AMD, injected into the lumbar subarachnoid space of cat in a dose of 1-1000 μg causes an interesting spinal cord disorder after a latent period. Distinctive changes occur during the latent period. (1) Neuronal and glial chromatin clumps and later coalesces into large dense masses (KOENIG & JACOBSON, 1966). (2) Coincidentally [^{14}C]adenine uptake into RNA is inhibited. (3) Protein synthesis, measured by [^{14}C]lysine incorporation, becomes depressed somewhat later, presumably owing to exhaustion of a small pool of short-lived (messenger?) RNA. (See Table 1). We postulate that AMD and related antibiotics on binding to DNA crosslink microfibrils of diffuse chromatin to form a dense chromatin devoid of template activity in the RNA polymerase reaction (KOENIG _et al_., 1967b).

TABLE 1. BIOCHEMICAL EFFECTS OF AMD ON CAT SPINAL CORD

	1 mg AMD				5 μg AMD	
	1 hr	4 hr	8 hr	24 hr	19 hr	40 hr
[^{14}C]Adenine uptake*	-65%	-72%	-97%	-	-61%	-
[^{14}C]Lysine uptake*	-	-29%	-81%	-96%	-61%	-95%
Nuclear RNA+	-	- 6%	-21%	-57%	-31%	-55%
Cytoplasmic RNA+	-	- 3%	-11%	-20%	- 9%	-18%
Soluble RNA+	-	0	-11%	-30%	-17%	-32%

*Relative specific activity   +mg RNA/g tissue

The spinal cord is still functional, e.g., cats walk, 16-24 hr after 1 mg of AMD when RNA and protein synthesis are virtually nil. Soon afterwards a spastic weakness of hindlimbs appears, followed by a flaccid arcflexic paralysis. This sequence evolves less rapidly after 1-5 μg of AMD. The pyrimidine analogue, 5-fluoroorotic acid (FO), causes a similar spinal cord disorder and impaired synthesis of RNA and protein (KOENIG, 1960), due partly to formation of analogue-substituted RNAs (KOENIG, 1967). Thus neuronal function is not immediately dependent upon RNA and protein synthesis, but cannot long endure if these biosynthetic activities are severely impaired.

KOENIG (1960). AMA Arch. Neurol., 2, 463. KOENIG & JACOBSON (1966). J. Cell Biol., 31, 61A. KOENIG & JACOBSON (1967). J. Neuropath. exper. Neurol., in press. KOENIG _et al_. (1967a) Trans. Am. Neurol. Ass'n., in press. KOENIG _et al_. (1967b). Fed. Proc., submitted.

[1]Assisted by grants from NIH (Nos. NB01456 and NB06838) and National Multiple Sclerosis Soc. (No. 372).

BIOCHEMICAL AND FUNCTIONAL CHANGES IN THE NERVOUS SYSTEM PRODUCED BY METHIONINE SULPHOXIMINE

Jaroslav Kolousek

National Institute of Health and Medical Research,
(Ghana Academy of Sciences)
Accra, Ghana*.

Methionine sulphoximine (MSI) produced metabolic changes in animals. An inhibition of [^{35}S]methionine (ME) incorporation in all tissues and organs was observed. In the CNS the highest inhibition was in the cerebellum. Incorporation of [^{35}S]ME into isolated brain protein was also inhibited. ME excretion in the urine was increased. The glutamine synthetase activity of brain was inhibited, depending on the dose. Glutamine formation in the brain was disturbed. Phosphorus metabolism in the body was influenced. Urinary excretion of exogenous [^{32}P] phosphates increased, but the excretion of endogenous phosphates decreased after a single dose of MSI. A strong excretion of ammonia nitrogen in the urine occurred. The ADP and ATP content of rat liver decreased; oxidative phosphorylation increased, but the intensity of respiration did not change after six hours. MSI did not change the intensity of respiration in the brain, but decreased esterification of inorganic phosphates. Oxygen under pressure normalized the uncoupling effect of MSI in the rat brain.

A maximal accumulation of [^{35}S]MSI in the CNS, especially in the hypothalamus, was observed six hours after a paroxysmal dose. [^{35}S]MSI was not incorporated into protein. The maximal content was in the non-protein fraction of the brain. Incorporation of [^{35}S]ME into the neurosecretory nuclei was inhibited. MSI damaged the neurosecretory nuclei of the hypothalamus. In rats the water intake and urine excretion decreased. Changes in higher nervous activity appeared. Food conditioned reflexes disappeared and cortical postural reflexes were inhibited. When the inhibition of reflex activity spread to cortical and certain subcortical regions tonic-clonic seizures appeared. ME reduced the susceptibility to seizures and normalized EEG records. Spontaneous preparoxysmal and evoked electrical activity of some amygdalar and hypothalamic nuclei of the brain showed a qualitative change of reactivity of neurohumoral systems after MSI (HREBÍCEK & KOLOUSEK).

*Address: P.O. Box 2848, Accra, Ghana,
West Africa.

ON BRAIN PROTEINS PARTICIPATING IN AMIDE NITROGEN METABOLISM

P.A. Kometiani, H.E. Klein and H.G. Gotsiridze

Institute of Physiology, Georgian Academy of Sciences, Tbilisi, U.S.S.R.

As is well known, the free carboxyl groups of brain proteins participate in the removal and release of ammonia. It has also been found that ammonia is an index of the functional state of the brain (VRBA, FOLBERGOVA & KANTUREK, 1958; KLEIN _et al._, 1963). However, it remains unknown which proteins take part in this process and where they are located.

Experiments with rats showed that nembutal increased the amidation of proteins, while camphor caused their deamidation. Stimulation of the oxidative metabolism of amino acids in brain slices incubated in the presence of glucose or keto acids was shown to produce significant changes in the amide nitrogen content of proteins. It was found also that the protein fraction extracted from TCA precipitates by acidified organic solvents was most active in the amidation and deamidation of proteins. Apparently these proteins are present in the fraction soluble in acidified ethanol, which was studied by GAITONDE (1961) and OTSUKI & GEIGER (1963). The amount of protein extractable by acidified organic solvents increased with rise of brain activity, as well as during incubation of brain preparations.

Changes in the rate of amidation of proteins in subcellular particles were studied also under conditions of increased ammonia production in the brain. This occurs mainly at the expense of proteins in the nuclear fraction. Decrease in the total protein amide nitrogen was found to be due to water-insoluble proteins, while the major part of the soluble proteins acts as an interceptor of ammonia.

Study of the electrophoretic (paper, agar) behaviour of soluble brain proteins in electrically stimulated rats, as also during incubation of brain preparations, showed that amidation of some soluble proteins was accompanied by deamidation of others, the two processes being reflected in an increase of the 'globulin' and the 'prealbumin' fractions, at the cost of proteins in the 'albumin' fractions. Fractionation experiments with thin-layer Sephadex G-200 showed that the changes observed in proteins are by no means confined to deamidation or amidation, being also attended by changes in molecular weight.

In a last series of experiments an attempt was made to separate the proteins participating in the metabolism of amide nitrogen by combining gel-filtration on Sephadex column with starch-gel electrophoresis. The data obtained suggest the possibility of a closer study of protein changes under different functional conditions of the organism.

GAITONDE, M.K. (1961). *J. Neurochem.*, *8*, 234. KLEIN, H.E. *et al*. (1963). *III Vsesojuz. Konfer. Biochimii Nerv. Systemi, Erevan*, p.193. OTSUKI, S. & GEIGER, A. (1963). *J. Neurochem.* *10*, 415. VRBA, R., FOLBERGOVA, J. & KANTUREK, V. (1958). *J. Neurochem.*, *2*, 187.

DETERMINATION OF Na^+ AND K^+ IN INDIVIDUAL INVERTEBRATE NEURONS

Z. Kometiani*, S. Hovmark and E. Giacobini

Department of Pharmacology, Karolinska Institutet, Stockholm.

A specially designed micro-flamephotometer (CARLSSON, GIACOBINI & HOVMARK, 1966) was used to study intracellular variations of Na^+ and K^+ in individual neurons from the stretch receptor organ of the crayfish.

The instrument measures the total emission from an individual nerve cell placed in a gas flame. The sample, on the tip of a 50 μ metal wire, is automatically introduced into the flame where excitation takes place. The emitted light of the spectral wave length is focussed on two photomultiplier tubes. From these, the output signals are displayed on a double beam oscilloscope and simultaneously electronically integrated. The integrated signals are fed into capacitor memories. The background signal is automatically subtracted from the sample signal and the result is registered on a potentiometric recorder.

The following points will be presented and discussed:

1) interference of biological material with the measurement of known concentrations of Na^+ and K^+.

2) interference of other ions with the measurement of Na^+ and K^+ in solutions and in biological samples.

3) intracellular variations of Na^+ and K^+ in isolated neurons under different conditions of activation and inhibition of ion transport.

CARLSSON, B., GIACOBINI, E. & HOVMARK, S. (1966). XII Scand. Congress of Physiol. (Turku), 1966. Acta Physiol. Scand., 68, 277.

*Present address: Institute of Physiology, Academy of Sciences, Tbilisi, USSR.

COMPOSITION PHOSPHOLIPIDIQUE DES DIVERSES TUMEURS CEREBRALES CHEZ L'HOMME

D. Kostic et Lj. Rakic

Les auteurs étudient la distribution qualitative et quantitative des différents phospholipides dan les tumeurs endocraniennes (Astrocytome gr. I-IV, Meningiomes bénins et malins et Carcinomes métastatiques).

Ils y ont séparé, par la chromatographie sur couche mince, les fractions suivantes: phosphatidyléthanolamine,
phosphatidylcholine,
sphingomyéline,
phosphatidylinositol
et phosphatidylsérine.

En analysant les résultats obtenus, ils constatent des différences notables dans les tumeurs en phospholipides totaux et dans la distribution des fractions isolées dans ces diverses tumeurs. Ainsi, dans les astrocytomes et les méningiomes malins, ils ont constaté une diminution des phospholipides. Parmi les fractions isolées la diminution porte spécialement sur les phosphatidyléthanolamines.

Les tumeurs métastatiques ont une composition nettement differente qui reflète sans doute la composition du tissu tumoral initial.

A COMPARATIVE STUDY OF PHOSPHOLIPIDS IN THE NERVOUS SYSTEM OF MOLLUSCS

E.M. Kreps

Institute of Evolutionary Physiology and Biochemistry, USSR Academy of Sciences, Leningrad

The phospholipid composition of nervous ganglia of the bivalve Unio crassa, the gastropod Helix pomatia and the cephalopods Octopus sp. and Ommastrephes sloanei pacificus were investigated.

The ganglia of cephalopods contained considerably more phospholipids than gastropod and bivalve ganglia. Especially rich in phospholipids were the optical ganglion of the squid Ommastrephes (4.5 mg P per g dry weight) and the cerebral ganglion of Octopus (5.6 mg P per g dry weight), i.e. of the same order as the brain of teleosts and amphibia. In the ganglia of the lower molluscs – the bivalve and the gastropod – no sphingomyelin nor diphosphoinositide could be detected.

In the optical ganglion of Octopus no sphingomyelin and no diphosphoinositide could be detected, whereas in its cerebral ganglion sphingomyelin, but no diphosphoinositide, was easily found; and it averaged 3.2% of the sum total of phospholipids.

In both the optical and the cerebral ganglia of the squid Ommastrephes sphingomyelin was present (4.6 and 3.5% respectively), as well as a phosphoinositde with a R_f similar to the R_f of diphosphoinositide, but differing from a typical diphosphoinositide in the shade of colour developing with Rodamin 6 G on the paper chromatogram.

Thus within the mollusc phylum the development of the central nervous system, considered from the point of view of its anatomical structure and functional manifestations as well as of its phospholipid composition, shows the same range of evolutionary changes as is seen in the nervous system over the whole animal kingdom.

COMPARATIVE RATES OF SYNTHESIS OF NAD IN NEURONAL AND GLIAL CELL NUCLEI

Masanori Kurokawa, Takahiko Kato and Hiroshi Inamura

Biochemistry Division, Institute of Brain Research,
Faculty of Medicine, University of Tokyo,
Tokyo.

Three types of highly purified cerebral nuclear preparations were isolated by an aqueous method. Identification of nuclei on a morphological basis showed that 47-53% and 72-83% of nuclei in the P4II and P_L preparations respectively consisted of neuronal nuclei. In contrast, 94-99% of nuclei in the P_S preparation consisted of non-neuronal nuclei, most likely of oligodendroglial origin. The nuclear preparations retained the activity of the DNA-dependent RNA polymerase system. The activity of ATP:NMN adenylyltransferase, as estimated by the formation of NAD from NMN and ATP in an appropriate medium, was 13.5, 3.4 and 6.8 mμmoles NAD/20 minutes/10^6 nuclei (38°) for the P4II, P_S and liver nuclei respectively. The available evidence indicates that the difference in the rate of NAD formation is not a consequence of cleavage of NAD, catalysed by NAD glycohydrolase. Dinucleotide nucleotidohydrolase was undetectable in the present nuclear preparations. On the assumption that the NMN adenylyltransferase activity is uniform in each neuronal nucleus and in each glial cell nucleus, the activity is estimated to be 24.1 and 2.5 mμmoles NAD/20 minutes/10^6 nuclei (38°) for neuronal nuclei and for glial cell nuclei respectively. A more detailed comparison of the enzyme activities among various cerebral nuclei, however, requires further differential isolation of respective glial cell nuclei, particularly of astroglial nuclei. The lowered value of NMN adenylyltransferase for glial cell nuclei implies a different turnover rate of NAD between the neuronal and glial cell systems, though the evaluation of this finding awaits elucidation of the general significance of NMN adenylyltransferase in the NAD metabolism in brain, particularly in connexion with the Preiss-Handler's pathway.

PURIFICATION AND PROPERTIES OF A-TYPE ESTERASES IN FOETAL AND ADULT RABBIT BRAIN

J.R. Lagnado and M.H. Hardy*

Department of Biochemistry, Bedford College (University of London), London.

It was previously found that zymograms of crude extracts of mammalian brain show two main A-type esterase fractions when α-naphthyl acetate is used as substrate. In each species examined (rat, rabbit, guinea-pig, pig and man), one of these fractions (A_2) is greatly decreased during early development when it is partly or completely replaced by the more anodic electrophoretic variant (A_1) prominent in adult brain (LAGNADO & HARDY, 1966).

In order to investigate the possible nature of this age-dependent alteration, the foetal and adult type isoenzyme variants were purified from crude rabbit brain and their chemical and kinetic properties were compared. The individual fractions were assayed by the naphthol-diazonium coupling method according to PILZ (1962) and their homogeneity was tested by the zymogram technique (LAGNADO, 1962).

A 20-50 fold purification of fractions A_1 and A_2 was achieved as follows. Crude aqueous extracts of foetal or rabbit brain were acidified to pH 5 with 2% acetic acid and A_1 and A_2 were precipitated from the resulting supernatant between 30-50% and between 50-75% saturation with $(NH_4)_2SO_4$, in that order. This was followed by chromatography on DEAE-cellulose columns equilibrated with 0.05 M-tris-HCl buffer (pH 7.6) containing 0.002 M-mercaptoethanol: fractions A_2 and A_1 were eluted stepwise with 0.1 and 0.2 M-KCl, in that order. Omission of mercaptoethanol resulted in the complete inactivation of all A-type esterases except A_1.

Furthermore, it was found that A_2 was much more susceptible to inhibition by organomercurial compounds and to heat inactivation than A_1. Both fractions A_1 and A_2 appear to behave as globular proteins of very similar molecular size (m.wt. 70-80,000) after chromatography on calibrated columns of Sephadex G-200 according to ANDREWS (1964). The possibility that the foetal and adult A-esterase variants differ mainly in their quaternary structure was therefore investigated and the results of experiments in progress will be described.

ANDREWS, P. (1964). Biochem. J., 91, 222. LAGNADO, J.R. (1962). Acta Neurol. Scand., 38, 63. LAGNADO, J.R. & HARDY, M.H. (1966). In Variation in Chemical Composition of the Nervous System. Edited by G.B. Ansell, p. 63. Pergamon Press, London. PILZ, W. (1962). Z. Physiol. Chem., 328, 1.

*M.R.C. Scholar.

BIOCHEMICAL PATTERNS IN ISOLATED MEMBRANES OF THE CNS

E.G. Lapetina, E.F. Soto* and E. De Robertis

Instituto de Anatomia General y Embriologia, Facultad de Medicina, Universidad de Buenos Aires, Argentina

While the general tendency is to center our knowledge of the cell membrane on the composition of myelin, on the unit membrane theory, and on a generalized model of molecular structure, there is increasing evidence in favour of a diversity of structure and function in biological membranes.

Various membranous structures, i.e. myelin, nerve ending membranes, junctional complexes, synaptic vesicles, mitochondria and microsomes have been isolated from the rat brain cortex and analyzed for total protein, proteolipid protein, total phospholipids, cholesterol, galactolipids, plasmalogens and gangliosides. A definite biochemical specialization reflected in the lipid composition of the different membranes will be demonstrated.

The presence of gangliosides in microsomes and in some of the isolated nerve ending membranes will be discussed in relation to the possible functional role of these acid glycolipids. The importance of proteolipids as structural constituents of myelin and other isolated membranes will be stressed.

Analysis of the molar ratios for the different lipids, proteolipid protein, amino acids and protein amino acid in the various types of membranes, show that each of them has a specific pattern of molecular organization.

Supported by Grants from the Consejo Nacional de Investigaciones Científicas y Técnicas and No. 456 from the Multiple Sclerosis Society.

*Presenting author

TISSUE SLICE EXPERIMENTS ILLUSTRATING INHIBITORS OF CEREBRAL OXIDATIVE METABOLISM IN URAEMIC AND HEPATIC COMA

P. Lascelles

Institute of Neurology, National Hospital, London W.C.1,England

HEYMAN (1951) and WECHSLER (1954) demonstrated a dminished oxygen uptake in vivo by the brain of patients suffering from uraemic and hepatic coma. The cause of this is at present unknown. One explanation is that cerebral oxygen uptake might be affected by abnormal accumulation of metabolites formed in these diseases. The hypothesis was investigated by studying the effect upon the oxygen uptake of rat brain, kidney and liver slices in vitro of metabolites known to accumulate in uraemic and hepatic coma.

Uraemia. Urea was found to inhibit the oxygen uptake of brain slices at concentrations occurring within the range found in human blood during disease, but there was no inhibition of oxygen uptake by kidney and liver slices.

Creatinine, creatine, indoxylsulphate, ethanolamine, p-cresol, guanidine and low concentrations of sodium were found to inhibit cerebral oxygen uptake also, but not at concentrations lying within the ranges occurring in disease. Guanidine also inhibited the oxygen uptake of kidney slices and p-cresol inhibited in liver slices.

Evidence is presented suggestive of mutual potentiation of inhibition in brain slices by metabolites present together in lower concentrations. Ultrafiltrates of serum from patients with uraemic coma did not inhibit the oxygen uptake of brain slices.

Hepatic Coma. Experiments confirmed the inhibitory effect of ammonium ions, skatole and indole on the oxygen uptake of brain slices. In addition indole was found to inhibit the oxygen uptake of kidney and liver slices, and skatole of liver slices. All metabolites exhibited this effect at concentrations lying outside the pathological range.

Human hepatic bile, and dihydroxy- bile salts (conjugated and unconjugated) inhibited cerebral oxygen consumption at concentrations occurring within the pathological range. Human hepatic bile appeared to inhibit liver slices and dihydroxy-bile salts inhibited in kidney slices. Mutual potentiation of inhibition in brain slices by these factors present together in lower concentrations was demonstrated. Ultrafiltrates of serum from patients with hepatic coma did not inhibit oxygen uptake in brain slices.

STUDIES ON LOW MOLECULAR WEIGHT PROTEINS PRESENT IN CEREBROSPINAL FLUID (C.S.F.) AND IN NORMAL AND PATHOLOGICAL URINES

E.C. Laterre, Ph. Walravens, A. Estas and J.F. Heremans

Laboratoire d'Analyses et de Recherches Chimiques
Clinique Universitaire de Neurologie Neurochirurgie
Louvain

A parallel study was undertaken on the proteins of C.S.F. and normal urines and on the proteins in urines from patients with tubular nephropathies of various natures, including renal transplantation. The authors' main interest was focussed upon the fractions which are non-identifiable in human plasma.

The two principal C.S.F. proteins which are apparently not derived from the blood plasma, are the γ_{trace} and β_{trace} proteins. These two proteins can be identified in normal urine and especially in urines collected from tubulopathic patients. Some other proteins, such as the β-microglobulin isolated by BERGGARD from urine, can also be detected in the C.S.F. The presence and the relative concentrations of some of these proteins in C.S.F. may depend on a simple mechanism of diffusion.

Such a mechanism cannot however explain the presence of all these proteins.

PURIFICATION OF BRAIN ACETYLCHOLINESTERASE

H. Claire Lawler

The Neurotoxicological Research Unit of the Department of Mental Hygiene of the State of New York at the Bronx State Hospital and New York State Psychiatric Institute

A brain acetylcholinesterase preparation from calves caudate nucleus was solubilized and subjected to a seven-fold purification by ammonium sulphate fractionation, density centrifugation and a second ammonium sulphate fractionation. The preparation could hydrolyse 145 μmoles of acetylcholine per mg of protein per hour. It differed from an electric eel tissue preparation in its stability and in its sensitivity to a number of reagents. Homogentisic acid, ascorbic acid, phenylmethane-sulphonyl fluoride and diphenylcarbamyl chloride decreased the enzymic activity. A concentration of 4×10^{-5} M-phenylmethanesulphonyl fluoride produced over 90 per cent inhibition, whereas at higher concentrations itdid not produce a measurable inhibition of the eel tissue acetylcholinesterase. The effect of inhibitors on the enzyme will be reported.

The instability of the purified fraction of brain acetylcholinesterase has made further purification difficult. Twenty per cent of its activity is lost per day at 6°. Amongst the series of compounds tested in order to reduce the activity loss it was found that the various forms of vitamin B_6 were effective. The stabilizing effect of pyridoxamine phosphate was investigated.

AMIDATION OF A SPECIFIC BRAIN PROTEIN FRACTION DURING CONVULSIVE ACTIVITY AND AMMONIA INTOXICATION

F.N. LeBaron

Department of Biochemistry, School of Medicine, University of New Mexico, Albuquerque, New Mexico

We have recently localized some of the amide binding in brain proteins to a specific fraction which has a relatively high content of glutamic and aspartic acids (LEBARON, 1966). In this presentation, we wish to report initial results bearing on the question of whether amidation of these proteins is related only to detoxication or is also related to functional activity of the brain. The protein fraction was prepared by a modification of the previously published procedure (MARTENSON & LEBARON, 1966). Brains of individual rats were homogenized in 2:1 $CHCl_3$:CH_3OH and the tissue residue washed exhaustively with $CHCl_3$:CH_3OH, acetone and water. Acid-soluble proteins were extracted from the resulting residue with 0.03 N-aqueous HCl, and a pH 7-insoluble fraction obtained by neutralization. This pH 7 precipitate was redissolved in 0.01 N-HCl and fractionated on a Bio-Gel P 150 Column. The protein in the first peak eluted by 0.01 N-HCl was subjected to analysis for total N, amide-N by the Conway Microdiffusion Method, and amino acid pattern using a Beckman-Spinco Model 120 C Amino Acid Analyzer. The resulting protein fraction was not homogeneous, since it could be separated into several components when subjected to electrophoresis on polyacrylamide gels.

Minimum normal values of amidation were obtained only when the brains were frozen in liquid N_2 immediately after sacrifice. Increases were found after short autolysis or intraperitoneal injection of 750 mg/kg doses of NH_4Cl, 20 to 40 minutes before sacrifice. Convulsions have been produced by various methods and the amidation of the proteins measured under a variety of conditions.

Supported by Grant NB 05923 from the National Institute of Neurological Disease and Blindness.

LEBARON, F.N. (1966). <u>Southwest Report</u>, <u>19</u> No.4, 27A
MARTENSON, R.E. & LEBARON, F.N. (1966). <u>J. Neurochem.</u>, <u>13</u>, 1469.

GANGLIOSIDES OF THE ADRENAL MEDULLA

Robert Ledeen, Kenneth Salsman and Maria Cabrera

The Saul R. Korey Department of Neurology and the Department of Biochemistry, Albert Einstein College of Medicine, New York.

The adrenal medulla is known to be derived embryologically from the neural crest and is a component of the autonomic nervous system. Its content of lipid-bound sialic acid was estimated to be 290 μg per gram of fresh tissue; this is somewhat greater than white matter of brain but approximately a third that of grey matter. However, 90% of the gangliosides were fast-moving (TLC) monosialo species, compared to brain which has predominantly di- and trisialo species. When expressed as moles of ganglioside per unit wet weight the medulla contains approximately 75% that of brain grey matter.

The major species consisted of two fast-moving bands on TLC. The faster contained N-acetylneuraminic and the slower N-glycolylneuraminic acid. Both contained glucose and galactose in equimolar ratio but lacked hexosamine. Neuraminidase reacted in each case to give a sialic acid-free glycolipid which migrated parallel to ceramide-lactoside. Approximately 10% of total ganglioside consisted of a group (5 detectable) which resembled brain gangliosides in their migratory rate and possession of hexosamine. Both types of sialic acid were present in these species and only part was liberated by neuraminidase.

The two major fractions contained similar fatty acid compositions, with 22:0 the most abundant (36%). Others present were 16:0 (13%), 18:0 (12%), 24:0 (14%), and 24:1 (11%). Lesser amounts of 16:1, 18:1, 20:0, and 23:0 were also detected. The slower-moving gangliosides, analysed as a group, showed approximately the same composition.

STUDIES OF PHOSPHOPEPTIDES IN ANIMAL TISSUES

Marc Ledig and Paul Mandel

Centre de Neurochimie du C.N.R.S.,
Institut de Chimie Biologique, Faculté de Médecine,
Strasbourg.

Twenty eight amidated phosphopeptides (PP), glycerophosphate, inositol-phosphate, glycerylphosphorylserine and non-phosphorylated compounds such as inositol, peptides and amino acids were isolated from FOLCH's phosphatidopeptide fraction by combining paper electrophoresis, paper and ion-exchange chromatography.

The PP were at first separated into two groups: one hydrosoluble, the other liposoluble. Liposoluble PP can be separated from phosphatides present in the extract, only by alkaline hydrolysis. One fraction of PP from both lipo- and hydrosoluble groups can be absorbed on activated charcoal.

PP are alkali stable; this characteristic differentiates them from phosphoproteins that release their phosphate as inorganic phosphate in alkaline media. PP are resistant to phosphatases and to proteolytic enzymes.

All the usual amino acids were found in PP with the exception of tryptophan and methionine. Serine and glycine are predominant. An important breakdown of serine which is phosphorylated and amidated occurs during acid hydrolysis. The amide and phosphate groups bound to serine may confer ion exchange properties on PP and this suggests that they may participate in transport phenomena.

A study of their subcellular distribution in liver and brain shows that they are associated with membranous structures and that the myelin sheath is very rich in PP. The increase of PP during postnatal brain development parallels the myelination process. The demonstration of PP in red blood cells and in bacteria is an indication of their wide distribution in nature. The rapid incorporation of ^{32}P into PP when compared to that of other phosphorylated compounds indicates that they have a high rate of turnover.

THE SOLUBILITY OF BRAIN PROTEINS IN CHLOROFORM-METHANOL MIXTURES

Marjorie B. Lees

McLean Hospital Research Laboratory, Belmont, Department of Biological Chemistry, Harvard Medical School, Boston, Mass., and Department of Pharmacology and Toxicology, Dartmouth Medical School, Hanover, New Hampshire.

Proteolipids are defined as lipoproteins which are soluble in chloroform-methanol. Although the white matter proteolipids originally isolated by FOLCH and LEES (1951) constitute a specific group of lipoproteins, other proteins may become chloroform-methanol soluble as a result of certain preparative procedures. The amount of chloroform-methanol soluble protein obtained from rat brain tissue homogenates which have been subjected to washing by repeated centrifugation or dialysis is two to six-fold greater than that obtained from untreated homogenates. The increase is a consequence of the removal of inorganic salts during the process of centrifugation or dialysis. Little or no increase is observed upon (a) the addition of the supernatant, the ashed supernatant or the diffusate to the washed homogenate before its extraction with chloroform-methanol, (b) the addition of inorganic salts to the washed homogenate (c) dialysis of the homogenate against a salt solution or (d) washing of the homogenate with a salt solution instead of with water.

The increase in the amount of chloroform-methanol soluble protein upon removal of inorganic salts is a general occurrence which has also been observed in homogenates of rat brain subcellular particles, liver and heart, and in bovine white matter. The essentially complete solubility of isolated myelin in chloroform-methanol can be explained on this basis, i.e., the removal of tissue ions in the course of the preparation of myelin results in an increase in the amount of myelin protein which is soluble in chloroform-methanol. In addition, the previously observed effect of sucrose in increasing the amount of chloroform-methanol soluble protein obtained from rat brain subcellular fraction is apparently superimposed upon an effect of the removal of salts.

FOLCH, J. & LEES, M.B. (1951). J.biol. Chem., 191, 807.

(Supported by U.S.P.H.S. Grants No. NB 04419 and 07136)

A BARRIER EFFECT AT THE LEVEL OF THE PIAL SURFACE OF THE BRAIN

E. Levin, G.J. Nogueira and P.M. Wright

Departamento de Investigaciones Experimentales
Hospital Israelita, Terrada 1164, Buenos Aires

In previous studies we have postulated that a barrier is operative between the CSF and the nervous parenchyma, both at the level of the ventricular cavities as well as at the external subarachnoid space. In the present report we examine further the role of the pia mater and its vessels in preventing substances from passing into the brain parenchyma when offered by the CSF route.

Plastic cylinders were implanted on the brain surface of cats, forming a 'cup' or isolated CSF compartment, from which the disappearance of different substances could be followed. The half-time of disappearance ($T\frac{1}{2}$) for amino acids actively metabolized by the nervous tissue was of the order of 400 min as compared with 50 min for ethyl alcohol, 130 min for ethyl-thiourea and 140 min for urea.

When the brain was stripped of the pia, and the plastic cup was on the naked cortex, the $T\frac{1}{2}$ for different substances increased, as if the pia were responsible for the disappearance of the compounds from the cup. Intravenous infusion of GABA, maintaining steady levels in the blood, prevented the disappearance of the amino acid from the pial cup; in reverse, when the pial cup was filled only with artificial CSF and GABA was infused into the blood, it appeared in the cup following first order kinetics.

Other experiments indicated that there is a direct communication between the CSF and the blood through the pial vessels, which consequently are another source for the formation and reabsorption of CSF.

In the nervous tissue underlying the cup, a small percentage of the substances under study was found only in the immediate vicinity of the pia. When the brain has its normal blood supply, the CSF route appears ineffective for providing materials to the brain and the pial meninx plays a barrier role, diverting substances from their free passage into the encephalic parenchyma.

UPTAKE OF EXOGENOUS MONOAMINES BY NEURONS OF THE CENTRAL NERVOUS SYSTEM

Walter Lichtensteiger and Heinrich Langemann

Department of Pharmacology,
University of Zürich, Zürich.

Exogenous monoamines are taken up from the circulation by groups of central neurons which have been identified in the mouse by means of the histochemical fluorescence method of FALCK and HILLARP. After depletion of the endogenous amines by reserpine and subsequent inhibition of monoamine oxidase by nialamide, noradrenaline is found to accumulate in normally catecholamine-containing neurons of the area postrema, of certain hypothalamic nuclei and of the olfactory bulb. In some of these nerve cell groups such an uptake can only be shown with high doses of noradrenaline in animals protected against its peripheral actions by an alpha-blocking agent, azapetine. Similar but quantitatively different results are obtained with other catecholamines. Exogenous catecholamines appear to be taken up from the circulation only by catecholamine-containing neurons situated completely or partly in regions outside the blood-brain barrier. The resulting distribution of fluorescence differs considerably from that observed after intraventricular injections of catecholamines. After parenteral injection, a selective demonstration of tubero-infundibular neurons terminating in the region of the primary plexus of the hypophyseal portal system is obtained.

Apart from possible physiological implications, the finding of catecholamine-containing neurons located outside the blood-brain barrier offers a model system for pharmacological *in vivo* experiments with substances which cannot cross the blood-brain barrier. Thus, it has been shown that 5-hydroxytryptamine is also taken up by normally catecholamine-containing neurons of the central nervous system which were previously depleted by reserpine. This uptake is influenced by drugs in a similar way to the uptake of catecholamines. Noradrenaline and l-dopa are further taken up by cells of the subfornical organ, where no catecholamines are found in untreated animals. This uptake differs from the aforementioned one in the high doses required to obtain an accumulation as well as in the relation between the dose administered and the resulting fluorescence intensity.

THE MECHANISM BY WHICH SIALIC ACID IN THE GANGLIOSIDES IS RENDERED IMMUNE TO SIALIDASE

K. Lipovac and A. Rosenberg

Columbia University Research Service, Goldwater Hospital, and the Department of Biochemistry, Columbia University, New York City

The original observation that all but one of the sialyl residues in the brain gangliosides are susceptible to hydrolysis by sialidase (ROSENBERG *et al.*, 1956) has received ample confirmation in many laboratories. The sialyl residue of hematoside [sialyl-(2→3)-galactosyl-(1→4)-glucosyl-(1→1)-ceramide] is susceptible to enzymic hydrolysis; but the sialyl residue of Tay-Sachs ganglioside [N-acetyl-galactosaminyl-(1→4)- and sialyl-(2→3)-galactosyl-(1→4)-glucosyl-(1→1)-ceramide] is not. Thus, substitution by N-acetylgalactosamine of the position next to that occupied by sialic acid on galactose appears to protect the sialic acid from attack by the enzyme. The protective effect of the N-acetylgalactosamine was assumed to result from its steric hindrance of the approach of the enzyme to the susceptible linkage of sialic acid. An inspection of molecular models shows that the N-acetyl-galactosaminyl group is not in a position to seriously block the glycosidic linkage of the sialyl residue.

We tested the inhibitory nature of N-acetyl-galactosamine in the ganglioside-sialidase and the sialyl lactose-sialidase systems. Our findings indicate an inhibitory mechanism other than steric.

Free N-acetyl-galactosamine, galactosamine, or glucosamine were added to the enzyme-substrate incubation medium. N-Acetyl-galactosamine and galactosamine were found to be strongly inhibitory. Kinetic studies showed that the inhibition was competitive ($K_I = 3.2 \times 10^{-4}$ M; K_M for gangliosides $= 3.9 \times 10^{-2}$ M) for both N-acetyl-galactosamine and galactosamine. Glucosamine was without inhibitory effect. The results of this study indicate that the gangliosides contain, adjacent to the inner sialic acid residue, a built-in inhibitor which competes for the active center of the enzyme (*C. vibrio* sialidase). The mechanism whereby such competition is effected will be discussed, and evidence for the enzyme-orienting properties of the lactosyl and the galactosyl N-acetylgalactosaminyl residues of the gangliosides will be presented.

ROSENBERG, A., HOWE, C. & CHARGAFF, E. (1956). *Nature, Lond.*, **177**, 234.

CYTOCHEMICAL STUDY OF STIMULATORY AND INHIBITORY NEURONS OF THE CEREBELLAR CORTEX

Z. Lodin, J. Muller, J. Faltin, M. Kazachasvili and P. Mares

Department of Cytochemistry and Cytophysiology of Nervous Cell, Institute of Physiology, Czechoslovak Academy of Sciences, Prague.

Recently relatively complete data were obtained on neuroanatomical and neurophysiological relations among single elements of the cerebellar cortex (ECCLES, SZENTAGOTHAI, HAMORI). Afferent impulses are transmitted by ascending fibres (directly) and further on by means of mossy fibres (indirectly) to the dendrites and soma of Purkinje cells. The Purkinje cell is the only efferent element of the whole system. The basket, stellate and Golgi cells are believed to play inhibitory roles by mediation of impulses to synapses localized on the soma and dendrites of Purkinje cells.

We tried to analyse by means of histochemical methods the possible metabolic differences between stimulatory and inhibitory systems. In order to obtain the exact relation of the differences in functional metabolism studies on neurons which were connected with the identical zone in vertical projection of the afferent impulse, were performed under resting as well as under different physiological conditions. We studied the changes of the metabolism of proteins, of RNA as well as changes of the activity of different enzymic systems (dehydrogenases, hydrolytic enzymes and esterases).

ATP AND PYRIDINE NUCLEOTIDE CONTENT OF LIGHT AND DARK ADAPTED VERTEBRATE RETINA

Richard N. Lolley and Elisabeth Racz

Neurochemistry Laboratories, Veterans Administration Hospital, Sepulveda, Calif., and Department of Anatomy, University of California School of Medicine, Los Angeles, Calif. 90024.

In vivo substrate concentrations of ATP and of the oxidized and reduced forms of NAD and NADP were determined in frog retina (Rana pipiens). Frogs were frozen rapidly in ether-dry ice, either after dark adaptation for 60 min or after a dark-light schedule of 30 min darkness, followed by 30 min of light. Tissue sections (20 μ in thickness) were cut from the frozen eye with a Cryocut apparatus, which maintained the tissue at -18 to -20°C. Tissue slices were lyophilized and stored in a desiccator. The retina was dissected from the frozen, dried slices, care being taken to separate the pigment epithelium from the retina. The dried retinal material was weighed on a microbalance and assayed for ATP and the pyridine nucleotides according to the methods of LOWRY *et al*. (1957) and LOWRY (1964). ATP was present at approximately 12 μmoles ATP/gm dry wt. NAD and NADH levels were 0.55 and 0.22 μmoles/gm dry wt., respectively. The NAD/NADH ratio varied from 2 to 3. NADP and NADPH values were observed to be 0.06 and 0.35 μmoles/gm dry wt., respectively. No significant differences were noted between light and dark adapted animals in any of these parameters. The *in vivo* concentration of ATP and the pyridine nucleotides of frog retina are very near those previously reported for rapidly frozen mammalian nervous tissue (LOLLEY, 1961; LOWRY, 1964). With this technique, rapidly changing metabolic substrates can be investigated in tissue which were previously inaccessible because of their anatomical location.

LOLLEY, R.N. (1961). *J. Neurochem.*, *7*, 289. LOWRY, O.H. (1964). *J. biol. Chem.*, *239*, 18. LOWRY, O.H. *et al*. (1957). *J. biol. Chem.*, *224*, 1047.

STRUCTURE AND BIOSYNTHESIS OF MYELIN IN THE MAMMALIAN CNS *IN VITRO*

C.E. Lumsden and K. Bradbury

Department of Pathology,
University of Leeds.

Some metabolic events occurring *in vitro* in myelinating cultures of rat cerebellum have been studied with the aid of DL-[3-^{14}C]serine, DL-[^{3}H(G)]serine, L-[4,5-^{3}H]leucine, [methyl-^{3}H]choline chloride and [7α-^{3}H]cholesterol which are all potential precursors in myelin formation.

Two 2 mm diam. slices of new-born rat cerebellum were explanted on to the surface of 7/8 x ½ in. collagen-coated coverslips, and maintained in roller tubes with 1.3 ml of feed comprising balanced salt solutions, together with sera and embryo extracts. The use of the Maximow assembly in conjunction with this system allowed detailed microscopical observation during the critical phases of myelination.

Initially the labelled substances were added to the feed during the first 28 days of culture (during this period myelination is particularly rapid) and then normal feed was given and cultures fixed at various times thereafter. The explants were sectioned and autoradiographs prepared. Grain counts over neurons and glia both yielded biphasic decay curves, but with important differences. Other cultures were exposed to labelled precursors for periods of 24 hours at various stages of culture maturation, and autoradiographs prepared. In the instance of serine utilization by Purkinje cells, the uptake was highest in the period immediately after explantation, when the synthesis of new axons is the pre-eminent morphological event. Neuroglial cells showed a similar, but lesser, peak, followed by a second, and larger peak at 30 days and corresponding to the period of active myelination. Effects of demyelination and remyelination were also studied. Exposure of cultures to labelled precursors for periods of a few minutes allowed a study of the initial sites of uptake and subsequent distribution of the label.

Thin-layer chromatography of the explants made possible the separation of 10 to 12 lipid components and comparisons at a biochemical level.

The respective contributions of neuron and glia to myelin will be discussed in the light of these data.

THE RELEASE OF UNESTERIFIED FATTY ACIDS IN BRAIN

G.G. Lunt and C.E. Rowe

Department of Physiological Chemistry,
University of Birmingham, Birmingham, 15.

Unesterified fatty acids were estimated in brain tissue. All levels of unesterified fatty acids are expressed as mg. methyl ester/g. tissue protein. The yields isolated from mouse brain by thin layer chromatography (mean $\pm$ S.D. for 6 determinations) were: brain frozen _in situ_ in liquid nitrogen, 1.06 $\pm$ 0.09; fresh brain, 0.89 $\pm$ 0.13; slices incubated at 0° in Krebs-Ringer bicarbonate for 2 hr., 2.99 $\pm$ 0.27; slices incubated at 37° for 2 hr. 3.21 $\pm$ 0.35.

A high proportion of unesterified fatty acid from brain frozen in liquid nitrogen (58%) and from slices incubated at 0° (37%), was accounted for by acids, the methyl esters of which, had gas-chromatographic properties corresponding to tetradecamonoenoic, hexadecatrienoic and heptadecatrienoic acids.

Comparatively high yields of unesterified fatty acid (2.6-7.0), were obtained from particulate subcellular fractions of whole rat brain prepared by the method of EICHBERG, WHITTAKER & DAWSON (1964). The combined yield from all fractions (3.2) was greater than that from fresh unfractionated brain (1.5) indicating that fatty acids had been released during preparation of tissue fractions.

From subcellular fractions of rat brain prepared in a mixture of saline and metabolic poisons (CLAYTON & ROWE, 1965) acids were obtained with the gas-chromatographic properties of tridecamonoenoic, pentadecamonoenoic, heptadecamonoenoic and nonadecamonoenoic acids, in addition to the unsaturated acids previously obtained from mouse brain mentioned above.

When total lipid extracted from slices of mouse brain, which had been incubated on gauze at 0° or 37°, was treated with diazomethane high yields (77.7 $\pm$ 25 (6) and 117.2 $\pm$ 37 (6) respectively) of methyl esters were obtained. The effect at 0° was partly reversible by re-incubation at 37° to give a value of 2.22 $\pm$ 0.12 (6).

We are grateful to the Medical Research Council of Great Britain for financial support of one of us (G.G.L.) and for a grant for expenses.

CLAYTON, P.A. & ROWE, C.E. (1965). _Biochem. J._, _97_, 39P.
EICHBERG, J., WHITTAKER, V.P. & DAWSON, R.M.C. (1964). _Biochem. J._, _92_, 91.

THE OCCURRENCE OF FLUORESCENT LIPIDS IN BRAIN

G.G. Lunt and C.E. Rowe

Department of Physiological Chemistry,
University of Birmingham, Birmingham, 15.

Fluorescent lipids were isolated from mouse brain slices which had been incubated at 0° (LUNT & ROWE, 1965). Chromatography on alumina (ROWE, 1964) revealed three components, A, B and C in order of increasing mobility, which fluoresced purple in ultra-violet light. Yields were in the range 0.1-0.9 mg/g tissue.

The components gave a purple colour characteristic of sterols when chromatographed on a thin layer of Kieselgel G followed by spraying with sulphuric acid and heating. Purple colours were obtained in chloroform solution with sulphuric acid and acetic anhydride (LIEBERMANN, 1885; BURCHARD, 1890). Component A gave a yellow colour in chloroform solution with sulphuric acid alone.

The components had ultra-violet absorption peaks in the range 202-213 mμ, (I), and absorption in the range 224-234 mμ, (II). The ratio, II/I, of absorptions increased in the components in the order A, B and C. Component C had, in addition, absorption in the range 259-274 mμ. Treatment with sodium borohydride lead to the appearance of absorption in the range 330-360 mμ in components A and B.

Methanolysis gave mixtures of methyl esters of fatty acids. Little or no methyl ester was obtained from the action of diazomethane. This is in contrast to the effect of diazomethane on the fluorescent compound isolated from rabbit brain by GEIGER & BAMPTON (1966).

Fluorescent components were not obtained from fresh tissue or from slices incubated at 37°. There was little incorporation of [1-^{14}C]acetate at 0°. Radioactive component C was obtained from slices which had been incubated with [1-^{14}C]-acetate at 37° and reincubated at 0°.

We are grateful to the Medical Research Council of Great Britain for financial support of one of us (G.G.L.), and for a grant for expenses.

BURCHARD, H. (1890). Chem. Zentr., 1, 25. GEIGER, W. & BAMPTON, A.E. (1966). Nature, Lond., 212, 510. LIEBERMANN, C. (1885). Ber. dtsch. chem. Ges., 18, 1803. LUNT, G.G. & ROWE, C.E. (1965). Biochem. J., 97, 38P. ROWE, C.E. (1964). Biochim. biophys. Acta., 84, 424.

AGENTS WITH SPECIFIC ACTIONS ON SODIUM AND POTASSIUM ION MOVEMENTS IN MAMMALIAN CEREBRAL TISSUES

Henry McIlwain

Department of Biochemistry, Institute of Psychiatry, Maudsley Hospital, London, S.E.5.

Effects of the following substances will be described and new data quoted.

1. Glutamate, which increased the passive entry of Na^+ in samples of cerebral neocortex incubated in vitro (BRADFORD & MCILWAIN, 1966). Subsequent or associated changes included alterations in tissue volume and K^+ content.

2. Protamine and certain alkylating agents, which inhibited reassimilation of K^+ after the tissue K had been diminished by electrical stimulation (MCILWAIN, WOODMAN & CUMMINS, 1961; EVANS & MCILWAIN, 1967).

3. Tetrodotoxin, which in extremely low concentrations inhibited the gain of Na^+ and loss of K^+ normally caused by electrical stimulation of the isolated tissue (MCILWAIN, 1967). A quantitative study showed that tetrodotoxin supplied at 1.5 mμmoles/g. tissue inhibited the Na^+ entry. It is to be noted that 5.3 mμmoles Na^+ enter each g. of tissue per stimulating pulse (JOANNY & MCILWAIN, 1963), and thus these findings contribute to specifying mechanisms of Na^+ entry during excitation.

BRADFORD, H. & MCILWAIN, H. (1966). J. Neurochem., 13, 1163.
EVANS, W.H. & MCILWAIN, H. (1967). J. Neurochem., 14, 35.
MCILWAIN, H (1967). J. Physiol. Submitted for publication.
MCILWAIN, H. & JOANNY, P. (1963). J. Neurochem., 10, 313.
MCILWAIN, H., WOODMAN, R.J. & CUMMINS, J.T. (1961). Biochem. J., 81, 79.

CHANGES IN THE CONCENTRATION OF THE γ_c-GLOBULIN OF BOVINE CEREBROSPINAL FLUID WITH DEVELOPMENT

C.F.C. MacPherson

Allan Memorial Institute of Psychiatry, McGill University, Montreal.

It was observed in an earlier study (1) of the immunoelectrophoretic patterns of bovine cerebrospinal fluid (CSF) that the γG-globulin appeared to increase as the animals grew to maturity while the γ_c-globulin decreased. The concentration of the protein has now been measured in the CSF and in serum withdrawn from newly-born, colostrum-deprived calves and from 4-month-old calves and 3-year-old steers, by an immunochemical method. The average γ_c-globulin concentration in bovine CSF withdrawn from newly-born, colostrum-deprived calves was found to be 1.4 mg per 100 ml. This level decreased sharply a few days after birth and the average value for 4-month-old animals was 0.44 mg per 100 ml. The CSF of 3-year-old steers contained 0.22 mg per 100 ml.

The γ_c-globulin concentration was 2.8 mg per 100 ml in pooled fetal serum, but it decreased by one-third at birth and then remained stationary up to 4 months of age. Bovine colostrum was found to have on the average 15 mg of γ_c-globulin per 100 ml.

The γ_c-globulin was distributed throughout the parenchyma of the brain and spinal cord. Approximately 0.01 mg of γ_c-globulin per gram of wet tissue was found in cerebral white matter and spinal cord while the cerebral grey matter and cerebellum contained about one-third this amount.

A COMPARISON OF THE METABOLISM OF BRAIN AND AN EXPERIMENTAL GLIAL NEOPLASM DURING ISCHEMIA

Howard S. Maker and Gerard M. Lehrer

Division of Neurochemistry, The Mount Sinai School of Medicine, New York

Enzyme activities and changes in the concentrations of several intermediates of carbohydrate metabolism during 10 minutes of ischemia have been compared in mouse brain, in an experimental ependymoblastoma growing in mouse brain and in the same neoplasm growing subcutaneously in mice. Sensitive, quantitative, fluorescence micro assays were used. The enzymes assayed were hexokinase, lactate, malate and glucose-6-phosphate dehydrogenases, NADP-linked isocitrate dehydrogenase and 3-glucuronidase. Enzyme levels were similar in neoplasm growing in brain and subcutaneously. A resemblance of the enzyme spectrum of the glial neoplasm and immature brain was noted. Hexokinase and malate dehydrogenase were lower and lactate, glucose-6-phosphate and NADP-linked isocitric dehydrogenases and 3-glucuronidase higher in the neoplasm than in adult mouse brain.

Data on early changes during ischemia will be reported for glycogen, glucose, glucose-6-phosphate, lactate, fructose-1, 6-diphosphate, ATP, ADP, AMP, and for reduced and oxidized pyridine nucleotides. Prior to ischemia the concentrations of glycogen, glucose-6-phosphate and ATP were higher and glucose and lactate lower in mouse brain than in glial neoplasm growing intracranially. The concentrations of ADP and AMP were similar in the two tissues prior to ischemia but the total adenosine phosphate was lower in the neoplasm. The response to ischemia of the brain and the intracranially and subcutaneously growing glial neoplasm will be discussed. There was considerable variation from tumor to tumor in concentrations of metabolites in the subcutaneously transplanted neoplasms both prior to and during ischemia. The data suggest that the differences in modes of metabolism of brain and tumors growing at the two sites may be primarily determined by blood supply rather than by intrinsic differences in metabolic capacities dependent upon differences in enzyme activities. The usefulness of the glioma as a model for the study of glial metabolism will be discussed.

SOME PROPERTIES OF THE LIMITING MEMBRANES OF SYNAPTOSOMES AND SYNAPTIC VESICLES

R.M. Marchbanks and V.P. Whittaker

Department of Biochemistry,
University of Cambridge, Cambridge.

The retention of typical soluble cytoplasmic constituents (K^+, lactate dehydrogenase) by synaptosomes (detached presynaptic nerve terminals) during a 4 hr isolation procedure (WHITTAKER, 1965) suggests that the limiting (external) membrane is completely sealed. The permeability of the membrane for ions and small molecules (e.g. K^+, Na^+, acetylcholine, D-[^{14}C]galactose) has been studied using a gel filtration procedure (MARCHBANKS, 1966; 1967). The rates of passive diffusion of K^+ and Na^+ are approximately the same as those found in intact nervous tissue. The volume accessible to them and to D-[^{14}C]galactose is approximately the same as that calculated from electron micrographs (CLEMENTI, WHITTAKER & SHERIDAN, 1966).

A substantial proportion of the lactate dehydrogenase is in an occluded state and is only rendered accessible to its substrate and cofactor by procedures which rupture the limiting membrane. Synaptosome preparations respire with glucose as the sole substrate suggesting that a full complement of the glycolytic enzymes are retained within the membrane.

In studies with acetylcholine, the existence of two compartments was confirmed. One, readily exchangeable with added radioactive acetylcholine, was identified with the synaptosome cytoplasm. The second, not readily exchangeable even in the presence of ATP and Mg^{++}, corresponds to the synaptic vesicles. The failure of added radioactive acetylcholine to exchange with vesicular acetylcholine suggests that either the vesicle membrane is impermeable to acetylcholine or the acetylcholine is in a chemically bound form in the vesicle.

This work was carried out while R.M.M. was a Junior Research Fellow of the U.K. Medical Research Council.

CLEMENTI, F., WHITTAKER, V.P. & SHERIDAN, M.N. (1966). Z. Zellforsch. 72, 126. MARCHBANKS, R.M. (1966). Biochem. J., 100, 65P. MARCHBANKS, R.M. (1967). Biochem. J. In press. WHITTAKER, V.P. (1965). Proc. Biophys. Mol. Biol., 15, 39.

CHOLINE ACETYLTRANSFERASE ACTIVITY IN SPINAL AND SYMPATHETIC GANGLIA OF CHICK EMBRYO

P.C. Marchisio

Department of Histology and General Embryology,
University of Turin Medical School, Turin, Italy.

By employing a new radiometric technique choline acetyltransferase activity was measured in single freeze-dried spinal and sympathetic ganglia dissected from chick embryos ranging from stage 24 H.H. (4th day) to hatching. In spinal ganglia the choline acetyltransferase activity rose and reached a first peak at stage 29 (6th day), then steadily fell down to the 11th. From the 12th day the enzyme activity rose again and a second peak was reached at the 16th day (stage 42). In the following days before hatching a slow but significant decrease in activity took place. The early activity peak occurred at the time of maximal mitotic rate; the later increase proceeded in parallel with the volume increase and with the appearance of a well developed endoplasmic reticulum within the neuroblasts. It seems likely that the behaviour of this enzyme is related to the increase in number and volume of the cells and to the development of new membranous structures in the neuroblasts. This hypothesis suggests a functional role for the choline acetyltransferase in the synthesis of membrane components, since its product, ACh, enhances the production of phosphatidic acid.

Choline acetyltransferase activity in developing sympathetic chains followed a similar pattern to that in the spinal ganglia; however its activity level was 50% higher than in spinal ganglia, and the first peak was reached later, i.e. at stage 34 (8th day). There are clearly some similarities between the choline acetyltransferase activity in the development of spinal and sympathetic ganglia. A similar interpretation, though tempting, is impossible at the present time since evidence is lacking on the mitotic rate, the differentiation, and the onset of synaptic structures within sympathetic ganglia.

FREE AMINO ACIDS OF HUMAN BRAIN DURING MATURATION

Y. Mardens, M. van Sande and A. Lowenthal

Department of Neurochemistry, Born-Bunge Foundation, Berchem-Antwerp

Free amino acids in human brain tissue were examined by paper and column chromatography. This study enabled us to observe regional differences between the cerebral cortex, the white matter and the basal ganglia.

In several human foetuses where the topography of the various tissue sections was taken in account, the distribution of amino acids, and in particular of cystathionine, changed during maturation.

SEROTONIN, NORADRENALINE AND SLEEP CHANGES DURING NITROGEN DEPRIVATION

J. Mark, Y. Godin, K. Kobayashi and P. Mandel

Centre de Neurochimie du C.N.R.S., Institut de Chimie Biologique, Faculté de Médecine, Strasbourg

MATSUMOTO and JOUVET (1964) suggested that there is a correlation between central nervous system (CNS) biogenic amines and the two states of sleep. We have already described a decrease of brain threonine and tyrosine levels during 'protein free diet' (MANDEL & MARK, 1965). The amount of other essential amino acids remained unchanged. In recent experiments, we found that the tryptophan level is also decreased during this diet. Since tyrosine and tryptophan are precursors of biogenic amines, we have carried out the present investigation in order to determine if the 'protein free diet' also modifies the biogenic amine levels of the CNS or behaviour. It appeared that CNS serotonin and noradrenaline significantly decrease after 8 weeks of 'protein free diet'. At the same time an increase in the daily sleep time was observed. The ratio of total sleep and paradoxical sleep remained unchanged.

MANDEL, P. & MARK, J. (1965). J. Neurochem., 12, 987.
MATSUMOTO, J. & JOUVET, M. (1964). C.R. Soc. Biol., 158, 2137.

THE METABOLISM OF THE BASIC PROTEINS OF RAT BRAIN DURING DEVELOPMENT

R.E. Martenson, M.K. Gaitonde and D. Richter

Medical Research Council Neuropsychiatric Research Unit, Carshalton, Surrey.

A highly basic protein rich in glycine and basic amino acids was isolated from acid extracts of white matter of beef brain; it induced experimental allergic encephalomyelitis (MARTENSON & LEBARON, 1966; KIES, ALVORD, MARTENSON & LEBARON, 1966). Acid extracts of rat brain prepared under similar conditions, when purified by carboxymethyl cellulose chromatography, yielded a basic protein fraction which was also found to be an inducer of experimental allergic encephalomyelitis. The disc electrophoretic technique was applied to obtain highly basic proteins directly from the original acid extracts. Acid extracts of rat brain showed five main components which migrated towards the cathode. In contrast, preparations of beef white or grey matter gave only two zones of basic proteins. The two fast-moving basic proteins in acid extracts were found to be a major fraction in adult rats; it constituted about 50% of the total proteins that migrated into the gel as basic proteins. These fast-moving basic proteins were absent, or present in amounts less than 2% of the highly basic proteins, in acid extracts of 10-day old rat brain. Unlike the acid extracts of adult rat brain, those of the 10-day old rats contained a relatively large amount (about one-third of the highly basic proteins) of a component of low mobility. The proteins in acid extracts of rat brain were labelled *in vivo* after administration of uniformly labelled [^{14}C]leucine or [^{14}C]lysine. This fraction constituted 31-49% of the total ^{14}C of the adult rat brain at 30 hr. after injection. The results indicated that acid extracts of rat brain also contain basic proteins other than those from myelin.

KIES, M.W., ALVORD, E.C., Jr., MARTENSON, R.E. & LEBARON, F.N. (1966). Science, 151, 821. MARTENSON, R.E. & LEBARON, F.N. (1966). J. Neurochem., 13, 1469.

EFFECT OF IONIC STRENGTH ON GLUTAMATE DECARBOXYLASE ACTIVITY OF MOUSE BRAIN MITOCHONDRIA

Guillermo H. Massieu* and Bertha G. Ortega

Instituto de Biologia, Universidad National Autonoma de México, México, D.F.

Glutamate decarboxylase (GAD) activity of mouse brain mitochondria, measured by the γ-aminobutyric acid production, was depressed by substrate excess. This effect was ascribed to the corresponding excess of NaCl resulting from the adjustment of the substrate solution (glutamic acid hydrochloride) to pH 6.3, with NaOH.

When NaCl, KCl, NH_4Cl, $MgCl_2$ or Na_2SO_4 were added in increasing proportions to an incubating-phosphate buffered medium (pH 6.3; glutamate 0.0062 M; pyridoxal phosphate, 1 to 2.5 μg per tube) to raise the ionic strength from 0.84 to 1.16, a remarkable inhibition of the GAD mitochondrial activity was observed. Such action tended to be more intense when $MgCl_2$ or Na_2SO_4 were added.

The salt effect described was not modified by the presence of pyridoxal phosphate excess or glutathione addition to the medium. The possible role of the fluctuation of the levels of some ions on the regulation of γ-aminobutyric acid level in certain brain areas is discussed.

*Present address: Instituto Politécnico Nacional, México, D.F.

THE EFFECT OF AN ELECTRIC FIELD ON CEPHALIN FILMS

Leopold May and Arvind Kamble

Department of Chemistry, The Catholic University of America, Washington, D.C. 20017

As part of a programme to understand the nature of the interaction between an electric field and the neuronal membrane, the effect of a direct current potential applied across cephalin films was studied using infra-red spectroscopy. Commercial preparations of sheep brain cephalin purified by treatment with ether and acetone were used to prepare the films. It was found that upon applying an electric field between zero and 10,000 volts per millimeter certain bands in the infra-red spectrum were altered in intensity. The intensity of the band was determined by measuring the area under the band. The ratio of the areas at various field strengths to the area when no voltage was applied for a film increased to a maximum and then decreased. The value of the field strength at which the ratio reached a maximum depended upon the thickness of the film. It was also observed that the intensities of four bands varied with the applied field. These bands can be related to both the rocking vibration of the $-(CH_2)_n-$ groups found in the hydrocarbon portion of the lipid and the band assigned to P-O-C stretching frequency. Interpretation of these results will be presented.

COMPARISON OF THE PROTEIN COMPOSITION OF MYELIN FROM DIFFERENT SPECIES

E. Mehl

Biochemical Division, Max-Planck-Institute of Psychiatry, Munich, Germany

Myelin was isolated (AUTILIO *et al.*, 1963) from the cerebral white matter of man, ox, rabbit, guinea pig, rat and chicken. In the case of carp, whole brain was used for isolation. The purity of myelin was determined by using electron microscopic controls and enzymic assays. The purpose of this work was to find out the nature of quantitative relationship(s) between classical proteolipid (LEES *et al.*, 1964) and basic protein (LAATSCH *et al.*, 1962) that participate in the molecular structure of myelin.

Freeze dried myelin (10 mg) were dissolved completely in 1 ml of a phenol/formic acid/water mixture, 14:3:3 (w/v/v) to the monomeric state of proteins and lipids. High resolution electrophoresis of myelin proteins on polyacrylamide gel was performed in this solvent instead of conventional buffers, similar to the technique of WORK (1965). After staining with Amidoblack, at least 10 distinct protein fractions were detectable without any residual proteins remaining at the starting zone. The R_F-values were related to the migration of cytochrome C which was used as a standard.

The myelin samples of all species studied contained the classical proteolipid (R_F 0.54) as the main constituent, as well as the basic protein (R_F 0.93), except for the carp, where the basic protein was replaced by two proteins (R_F 0.83 and R_F 1.12). In the myelin from rat brain a second basic protein (R_F 1.08) was present. Other additional myelin proteins amounted to less than 15% of the total protein content. If we assume that the main constituents have identical binding capacity for Amidoblack, then 1.0 g proteolipid is accompanied by 0.70 g basic protein in myelin from man, ox, rabbit, guinea pig and chicken.

AUTILIO, L., NORTON, W.T. & TERRY, J. (1964). *J. Neurochem.*, *11*, 17. LAATSCH, R.H., KIES, M.W., GORDON, S. & ALVORD, Jr., E.C. (1962). *J. Exptl. Med.*, *115*, 777. LEES, M., CARR, S. & FOLCH, J. (1964). *Biochim. Biophys. Acta*, *84*, 464. WORK, T.S. (1964). *J. Mol. Biol.*, *10*, 544.

A POSSIBLE ROLE OF STRANDIN IN THE REGULATION OF CATECHOLAMINE TRANSPORT ACROSS MEMBRANES

Herbert L. Meltzer and Gerald Cohen

Departments of Biochemistry and Psychiatry,
New York State Psychiatric Institute and Columbia University.

In previous experiments (MELTZER, 1966) strandin solutions have been shown to be capable of accumulation of excess salt and of preferential cation uptake, when equilibrated across a dialysis membrane with various mixtures of sodium and potassium chloride. In the current experiments, when solutions of adrenaline, noradrenaline and strandin were subjected to equilibrium dialysis against distilled water, it was always observed that the catecholamine concentration on the strandin side ("inside") of the membrane exceeded the concentration on the opposite side ("outside"). When the "inside" solution was subjected to ultracentrifugation at 100,000 g for 3 hours to remove the strandin, the catecholamines were found entirely in the supernatant. Thus the excess of catecholamine in the "inside" solution did not arise because of simple ionic binding by the strandin. A solvent (water)-solute (strandin) interaction can be postulated as a reasonable mechanism for the accumulation of excess adrenaline and noradrenaline.

If, as is generally believed, strandins (or gangliosides) are components of nerve cell membranes, the observations cited above are relevant to the possible regulation of catecholamine transport across such membranes. It is not unreasonable to speculate that such solvent-solute interactions may be altered by other solutes, and that the passage of catecholamines through such interacting systems, localized at cell surfaces, may be controlled by the capacity of the new systems to accumulate excess catecholamines. That is to say, the rate of catecholamine transport may be a function of the extent and nature of the solute-solvent interaction.

(Supported in part by Grants No. 54-13 from the National Multiple Sclerosis Society and No. HE 01045 from the U.S. Public Health Service.)

MELTZER, H.L. (1966). J. Neurochem., 13, 423, 429.

ENZYMIC FUNCTIONS RELATED TO REDUCED PYRIDINE NUCLEOTIDE REGULATION IN BRAIN

S.A. Millard and E.M. Gal

Division of Neurochemistry, Department of Psychiatry, College of Medicine, University of Iowa, Iowa City, Iowa.

The mitochondrial localization of cerebral tryptophan and tyrosine hydroxylases as well as many of the diaphorases involved in reduced pyridine nucleotide generation led us to explore the biochemical interrelationship of some of these enzymes. Detailed studies have revealed in all species examined that the relative enzyme content of lipoyl dehydrogenase (LipDH) and tryptophan hydroxylase of cerebral mitochondria is area dependent. The mitochondria isolated from the cerebellum showed much higher lipoyl dehydrogenase activity than mitochondria of other brain areas; yet the cerebellar mitochondria displayed no tryptophan hydroxylase activity. The appearance of LipDH activity was found to be functionally related to the development of the brain. The values from the cerebellum from fetal to adult brain seemed to indicate that the increased LipDH activity is a function of an age-dependent increase in motor activity. The LipDH activity in the cortical and subcortical areas appeared to be almost equal whereas most of the tryptophan hydroxylase activity resided in the mitochondria from subcortical areas. Upon examination of discrete areas of rabbit brain an almost three-fold LipDH activity of the pineal body and the hypophysis was noted when compared to values from other areas.

Following purification of these enzymes from brain mitochondria and having earlier established the inhibition of liver hydroxylases by FAD at 10^{-4} M level, we were led to study the competition of FAD and reduced pteridines in these enzyme systems. Results of these studies will also be presented.

ENZYMIC PATTERN CHANGES IN CAT NEOCORTEX PRODUCED BY MESCALINE EPILEPTOGENIC FOCUS

Nella Mison-Crighel and G. Badiu

Neurological Institute of the R.S.R. Academy, Bucharest.

Determinations of succinate dehydrogenase, fumarate hydratase, aconitate hydratase and of some NAD- and NADP-linked dehydrogenases and aminotransferases were done on neocortical tissue taken at various times from control animals and from cats with epileptogenic foci produced by topical application of mescaline to the left ectosylvian gyrus. Enzymic activity was estimated in crude extracts of brain homogenates by spectrophotometric methods.

Electrocorticographic recording showed spike activity at the focus, appearing 2-3 min after topical application of mescaline, and general electrical changes both before and at 30 min after the onset of the focus. Mescaline-induced spike foci brought about changes in the pattern of enzymic activity, not only at the focus, but throughout the whole neocortex, the extent of these changes being, however, different in the various areas studied.

Before the onset of the spike focus, 2 min after the application of mescaline, succinate dehydrogenase activity increased significantly. At ten minutes, malate dehydrogenase activity decreased significantly at the focus only, while glutamate dehydrogenase decreased in both ectosylvian gyri. At 30 min after topical application of mescaline glutamate dehydrogenase activity was still low in the same areas, while malate dehydrogenase activity was increased in all areas investigated. The changes in enzymic pattern which accompany variations in excitability may account for some of the general metabolic disturbances induced by a spike focus.

PREPARATION AND TURNOVER STUDIES OF FOUR BRAIN PROTEINS

B.W. Moore, A. Pena-Ramos and V.J. Perez

Washington University School of Medicine, St. Louis, Missouri.

A soluble protein (S-100) has been purified from brain and has been shown to be unique to the nervous system and to be detectable in brains of all vertebrates studied in an immunologically similar form. We report now the preparation of 3 new soluble proteins from brain which appear (on the basis of DEAE-cellulose X starch gel two-dimensional patterns) to be absent from liver. The molecular weight of one of these was 30,000 (S-100 = 30,000) and of the others 60,000. An attempt is being made to prepare antiserum to these three as has already been done for S-100. The incorporation of [^{3}H]leucine into these 4 proteins is being studied in rat brain at various intervals after injection of isotope. The proteins are isolated for counting by cutting out stained bands from the gels after the DEAE-cellulose X starch gel fractionation; the gel sections containing the proteins are burned to H_2O which is counted. The S-100 protein was labelled to a smaller extent at short time intervals than the average labelling of brain soluble proteins.

DIRECT ACTION OF DEUTERIUM ON ISOLATED NERVOUS TISSUES IN ORGANIZED CULTURE

Margaret R. Murray and Helena H. Benitez

Columbia University, New York.

D_2O administered to mice in their drinking-water produces "sympathicomimetic" symptoms at sublethal replacement-levels (BARBOUR, 1937). When replacement of normal body water is continued beyond about 30% the animals die, and autopsy reveals destructive changes in several organs including the submandibular gland (BACHNER _et_ _al_., 1964); but the nervous system appears grossly normal.

Our enquiry was designed to find out whether D_2O exerts a direct effect on the sympathetic nervous system and its central ramifications, or whether an abnormal discharge of Nerve Growth Factor (NGF) from the damaged submandibular ducts might have been the mediating agent in the hyper-excitability of BARBOUR'S deuterated mice. Autonomic ganglia from embryonic chicks and rodents were maintained as organized, developing cultures for 2 months or more, during which time they were continuously exposed to D_2O incorporated into their feeding medium as 5, 12.5 or 25%. Cultures exposed to 33% D_2O died in about 2 weeks. Controls received portions of balanced isotonic salt solution, normal water or NGF. This series contained 440 cultures. Light-microscopic observations of the living cell communities indicated that deuteration within viable limits (up to 25%) accelerated and increased the growth of sympathetic neurons and favoured their repeated subdivision as a very large size is attained. The total neuronal population was thus increased by a factor of 2 or 3, and the progeny — clusters of small, neuroblast-like cells — enlarged and became multipolar, apparently recapitulating early stages of neurogenesis. In electron micrographs (by Dr. Edmund Masurovsky) the deuterated neurons and satellite cells displayed an extraordinary abundance of granular and fibrillar elements in both nucleus and cytoplasm; mitochondria appeared unusually dense, but not distorted.

Preliminary findings from organized cultures of developing cerebral cortex, cerebellum and hypothalamus indicate that these central nervous tissues are accelerated in growth and maturation by D_2O. Neurons are larger, both neurons and glia suffer fewer population losses than normally in culture; myelin sheaths develop earlier and in greater quantity and extent. For these areas of the brain, optimal D_2O ratios vary between 10 and 25% of the total medium. All these developing nervous tissues are able to tolerate substantial amounts of D_2O in their ambient medium for periods of weeks or months while remaining within normal structural and functional limits. In this partial-replacement situation, D_2O seems to encourage growth and maturation in both CNS and SNS; in the latter it also activates neuron reproduction.

A STUDY ON THE PRECURSOR FOR ACETYLCHOLINE SYNTHESIS IN THE NERVOUS TISSUE

R. Nakamura and S.-C. Cheng

New York State Research Institute for Neurochemistry and Drug Addiction, Ward's Island, New York.

We have demonstrated that in the isolated lobster nerves, acetate and, to a lesser extent, pyruvate and citrate could be incorporated into the acetyl moiety of acetylcholine (ACh). However, the radioactivity in C-5 of added glutamate was not found in this acetyl group. Using slices from both guinea pig cerebral cortex and rat striatum, we found among various substrates tested [2-^{14}C]pyruvate, [1-^{14}C]pyruvate, [1-^{14}C]-glutamate, [5-^{14}C]glutamate, [1,5-^{14}C]citrate and [2-^{14}C]acetate only [2-^{14}C]pyruvate was incorporated in significant amounts into the acetyl moiety of ACh; and [5-^{14}C]glutamate and [2-^{14}C]acetate contributed far less radioactivity. The other substrates failed to serve as a precursor for this acetyl group. Since the specific activity of glutamate, aspartate, citrate and lactate from [2-^{14}C]acetate and [1,5-^{14}C]citrate experiments were considerably lower than those from the [2-^{14}C]-pyruvate experiment, it is assumed that the permeability of these two substrates into the brain slices was greatly limited.

The fixation of C-5 of glutamate into the acetyl group of ACh confirmed the presence of some backwards reactions of the Krebs cycle in the brain slices. It implies that CO_2 fixation occurred at the oxalosuccinate level in these slices. However, these backwards reactions failed to contribute significantly to the synthesis of ACh from glutamate. A similar conclusion had already been reported in the lobster nerves.

In rat striatum slices, the specific activity of the synthesized ACh reached almost one-third of that of added [2-^{14}C]pyruvate. It suggests that acetyl coenzyme A is rapidly synthesized from pyruvate in the brain slices. In contrast, [2-^{14}C]acetate which was the best substrate for the lobster nerve, the specific activity of the synthesized ACh reached only one-tenth of that of added acetate. In the latter tissue with [2-^{14}C]pyruvate less than two per cent incorporation was found. Comparing the lobster nerve with brain slices, both of which synthesize ACh at a significant rate, a large metabolic difference was demonstrated.

(Supported in part by P.H.S. Grants NB-557 and NB-5869).

RAT BRAIN MYELIN: COMPOSITIONAL CHANGES DURING DEVELOPMENT

William T. Norton, Shirley E. Poduslo and Kunihiko Suzuki

The Saul R. Korey Department of Neurology, Albert Einstein College of Medicine, New York, N.Y. 10461

A method has been developed for the isolation of myelin of the same degree of purity from rats of different ages. The successful elimination of microsomal contamination from young myelin required scaling the method to have a constant amount of brain per unit volume during centrifugation, rather than a constant amount of myelin per unit volume. Myelin prepared in this way from animals ranging from 15 days to 14 months had 0.13-0.26% nucleic acid, representing 0.15-2.0% of the total homogenate value; a total ATPase activity of 0.3-0.8 μmole P/mg/hr, representing 0.2-0.6% of the total homogenate value; and 35-68 μg ganglioside NANA per 100 mg., representing 0.2-4.5% of the total homogenate value. Although the amount of these various constituents recovered in the myelin fraction increased with increasing age (and myelin yield), the value of each of these criteria was constant per unit weight of myelin at all ages. The amount of myelin recovered per brain (4, 12, 24, 41, 56, 64 and 91 mg. at 15, 20, 30, 60, 144, 190 and 425 days of age, respectively) was a linear function of the logarithm of age, throughout the whole range of ages studied. The curve extrapolated to zero myelin at 12 days of age. At all ages the isolated myelin accounted for approximately 40% of the total brain galactolipid, indicating we were obtaining a constant fraction of the total myelin content.

Although the total lipid content of the myelin was relatively constant, the percentages of the individual lipids varied with age. These variations were most significant in cerebrosides, sulphatides, lecithin and desmosterol. The total galactolipid increased from 21.4% of the total lipid at 15 days to about 30% at 60 days and then levelled off. The lecithin decreased from 16.4% at 15 days to 11% at 5 months and then levelled off. During the same period desmosterol decreased from 3.0% of the total sterol to 0.3%. None of the "lower phase" lipids showed significant changes with age.

Although the ganglioside NANA level was constant at all ages, the ganglioside pattern in myelin changed dramatically. From 6 months onward the major mono-sialoganglioside, G4 (SVENNERHOLM'S G_{M1}) accounted for 70-85% of the total myelin NANA. At 14 months the myelin fraction contained 17% of the total G4 of the whole brain, although it had only 4.5% of total brain NANA.

This study essentially corroborates the work of HORROCKS et al. (1966) on mouse myelin, and that of CUZNER and DAVISON (1966) on rat myelin, and at the same time eliminates a major criticism of these earlier studies that the variations in composition with age could be attributed to varying degrees of myelin purity.

This work was supported by U.S. Public Health Service Grants NB-02476, NB-01006 and NB-03356 and Grant No. R-160-65C from the United Cerebral Palsy Foundation.

HORROCKS *et al*. (1966). In *Variation in Chemical Composition of the Nervous System*. Edited by G.B. Ansell, p.46. Pergamon, Oxford. CUZNER & DAVISON (1966). In *Variation in Chemical Composition of the Nervous System*. Edited by G.B. Ansell, p.22. Pergamon, Oxford.

RATES OF PROTEIN SYNTHESIS AND CATABOLISM IN THE DEVELOPING RAT BRAIN

S.S. Oja

Institute of Physiology, University of Helsinki, Finland.

In the present study the rates of synthesis and catabolism of brain proteins were determined at different ages during the postnatal development of the rat, using [^{3}H]tyrosine as a label. The calculations were effected by a digital computer, using differential equations derived from a simple compartmental model. Simultaneously also the influx and efflux rates of tyrosine in the brain were determined.

The rated influx of tyrosine into the brain in the newborn was 0.8-1.2 μg/min/g of tissue. It increased about fourfold during the first two weeks after birth but then decreased again, at least as much until the adult age. The efflux showed a similar pattern of change, but the rate was only a third or half the rate of influx in the newborn, owing to the rapid growth of the brain. In the newborn the brain used 0.7-0.9 μg of tyrosine/g tissue/min for the synthesis of protein, while only a third of this amount was liberated as a result of its breakdown. In the adult rat both these rates were equal: 0.4-0.6 μg/min/g of tissue. The biological half-life times of free tyrosine in the brain were 20-30 minutes in the newborn; half as long in the adult, but only 4-5 minutes at the age of 14 days. The mean half-life times of brain protein varied from 4 to 9 days in the newborn, from 2 to 3 days at 14 days' age and from 4 to 6 days in the adult.

The efflux and influx of tyrosine in the brain reflect partly the changes in the protein metabolism of the brain. The results suggest further that the rate of brain growth is regulated by the rates of synthesis and catabolism of its protein.

BIOCHEMICAL AND BEHAVIOURAL CHANGES ASSOCIATED WITH FOCAL DEUTERON IRRADIATION OF THE MOUSE BRAIN

J.M. Ordy, T. Samorajski, W. Zeman and H.J. Curtis

Cleveland Psychiatric Institute, Cleveland, Ohio

Based primarily upon whole body or head irradiation and qualitative morphological criteria, most earlier studies suggested that the central nervous system of adult mammals may be highly resistant to ionizing radiation. More recently, a number of studies have demonstrated behavioural arousal as well as electrophysiological changes in the brain even after moderate or low levels of whole body or head irradiation. However, behavioural changes have not been related to specific biochemical changes in the brain. The aims of the present investigation were to study changes in protein, DNA, RNA, and neurotransmitter content in relation to changes in behaviour after focal deuteron irradiation of the mouse brain.

The subjects for this study were 80 C57BL/10 adult female mice, including 20 sham-irradiated controls. The focal irradiation of the brain was performed with a 20-Mev, 9 x 5 mm deuteron beam, at 0, 500, 5000, and 10,000 rads by the 60-inch Brookhaven cyclotron. The beam was directed through an aperture to the dorsal surface of the cerebrum and cerebellum. The mice were tested in a behavioural test battery before and after irradiation. They were sacrificed 15 months after the irradiation for biochemical evaluations of structural and functional changes in the brain.

Depending on the dose level, statistically significant differences were observed beginning 1 month after irradiation in spontaneous locomotor activity and maze exploration in the irradiated mice in relation to the behaviour of the sham-irradiated controls. There was a dose-dependent reduction in brain weight, and consequently in total protein, DNA, and RNA content, but not in their concentration. Similarly, there was a reduction in total cholinergic (ACh) enzyme activity as well as in 5-HT and NE levels.

From the statistically significant decreases in total protein, RNA, and DNA content and the non-significant changes in concentration and RNA/DNA, protein/DNA, protein/RNA ratios, it was concluded that the effects of deuteron irradiation resulted in a general loss of cells rather than in significant changes in specific cellular components. Similarly, from the significant effects on the neurotransmitter content and the non-significant alteration in concentration, it was also concluded that irradiation effects resulted primarily in a general reduction of neurotransmitters in the brain. The significant behavioural changes observed after deuteron irradiation of the brain indicated that behavioural changes can be associated ultimately with specific biochemical changes in the brain.

Supported in part by NIH Grant NB-3021.

ADENOSINETRIPHOSPHATASE DU NERF NORMAL, RÉGÉNÉRÉ ET DÉGÉNÉRÉ

A.V. Palladin et O.M. Rojmanova

Institut de Biochimie de l'Academie des Sciences de la R.S.S. de l'Ukraine, Kiev.

La neurochimie a enrégistré ces dérnières années des succés importants, obtenus en tout premier lieu grace aux études entreprises sur le systeme nerveux central. Les études concernant les processus biochimiques, qui se déroulent dans les nerfs, sont encore peu nombreuses. Pourtant les résultats déjà obtenus dans ce domaine montrant que l'activité métabolique dans les nerfs est assez importante.

Les nerfs périphériques possédent une propriété biologique très importante: la capacité de régénérer. Tout ceci nous a incité à entreprendre l'étude de l'adénosinetriphosphatase (cet enzyme qui joue un role très important dans le métabolisme énergetique) du nerf normal, régénéré et dégénéré. Nous avons étudié l'adénosinetriphosphatase activée par le magnesium (Mg) dans les homogénats et les fractions subcellulaires (mitochondriales et microsomales) des nerfs: le nerf sciatique, qui est péripherique, et le nerf optique, qui est un nerf cranien. Ces études ont été réalisées sur les nerfs de lapins, de chats et de chiens.

Nous avons pu mettre en évidence que l'activité de l'enzyme dans l'homogénat du nerf sciatique était assez forte et supérieure à l'activité de l'ATP-ase dans sa fraction subcellulaire mitochondriale; il y a lieu de noter que dans le cerveau l'activité de l'ATP-ase dans l'homogénat est sensiblement moins forte que dans les mitochondriales. La plus forte activité de l'ATP-ase se trouve dans la fraction des microsomes, ce qui présente de l'interêt du fait beaucoup lui attribue un rôle particulior dans la fonction physiologique de la cellule nerveuse: la conduction de l'influx nerveux. Dans le nerf optique l'activité de l'ATP-ase de l'homogénat et de toutes les fractions subcellulaires est supérieure à l'activité de l'enzyme de l'homogenat et des fractiones subcellulaires correspondantes du nerf sciatique. L'étude de l'activité de l'ATP-ase du nerf sciatique du chat, en état de régénération après séction, et du nerf ayant subi une dégénération de Waller complete, a montré que dans la partie distale du nerf en cours de régénération l'activité de l'ATP-ase est fortement augmentée, étant de 13 à 18 fois supérieure à l'activité du nerf de contrôle. Dans la partie proximale du nerf séctionne l'activité de l'ATP-ase reste, en général, au niveau de l'activité du nerf de contrôle.

Dans les parties proximale et distale des nerfs régénérés et dégénérés l'activité de l'ATP-ase était la plus forte dans la fraction microsomale; l'activité relative de l'ATP-ase (en % de l'activité de la fraction correspondante du nerf de contrôle) était la plus grande dans la fraction des mitochondries, où l'ATP-ase subit pendant la régénération les changements les plus profonds; il semble que ce ne soit pas la même ATP-ase qui intervient dans la fonction de transport des cations et qui est localisée dans la fraction des microsomes.

A l'aide de la methode histochimique de WACHTEIN et de MEISSEL nous avons constaté que dans le nerf normal l'activité maximale de l'ATP-ase est localisée dans les cellules de SCHWANN; elle peut également être mise en évidence dans l'axone, mais pas dans le gaine de myeline. Dans la partie distale des nerf régénérés et dégénérés la forte activité de l'ATP-ase est localisée dans les cellules de SCHWANN qui proliferent.

CHANGES IN THE ACTIVITY OF AMINO ACID: tRNA-LIGASES IN THE MOUSE BRAIN DURING POSTNATAL DEVELOPMENT

R.S. Piha

Department of Biochemistry, University of Oulu, Oulu, Finland

The incorporation of amino acids into cerebral proteins is more rapid in the newborn than in the adult (GAITONDE & RICHTER, 1956; PANTCHENKO, 1958; PIHA & UUSITALO, 1964; GELBER *et al*., 1964). To study whether there is any parallelism between the rate of incorporation and the activity of the amino acid activating enzymes in the developing brain, the activities of amino acid:tRNA-ligases were determined in the mouse brain during postnatal development.

The activity of amino acid:tRNA-ligases of whole brain was determined by an ATP-^{32}PP exchange method which included a quick separation of ATP and PP by paper chromatography (PIHA *et al*., 1966).

Of the different cellular subfractions the soluble fraction showed the most rapid ATP-PP exchange in the presence of most of the amino acids studied. The microsomal fraction was active when leucine, lysine, threonine and methionine were present and even more active than the soluble fraction in the case of phenylalanine. The activities of the 15 different amino acid:tRNA-ligases studied varied to a large extent. The rate of activation of leucine was highest. No measurable activity was found for the arginine or glutamic acid:tRNA-ligases.

The activity of most of the amino acid:tRNA-ligases decreased gradually in the developing brain. The minimum activity, on an average 62% of the initial value, was reached at the age of 12-14 days. After that the activity increased and at the end of the third week it attained the same level as at birth, remaining constant until adult age. These findings may be compared with those of JOHNSON and LUTTGES (1966) who found that the incorporation of valine into proteins of the mouse brain decreased steadily from birth to the 12th day and then remained unchanged.

The results suggest that during the rapid growth of the brain a parallelism exists between the activity of amino acid: tRNA-ligases and the rate of protein biosynthesis, while no corresponding parallelism can be found at the adult age.

GAITONDE, M.K. & RICHTER, D. (1956). Proc. Roy. Soc. B, 145, 83. GELBER, S., CAMPBELL, P.L., DEIBLER, G.E. & SOKOLOFF, L. (1964). J. Neurochem., 11, 221. JOHNSON, T.C. & LUTTGES, M.W. (1966). J. Neurochem., 13, 545. PANTCHENKO, L.F. (1958). J. Physiol. USSR., 44, 243. PIHA, R.S., AIRAS, R.K. & AARI, L.I. (1966). Suomen Kemistilehti B 39, 204. PIHA, R.S. & UUSITALO, A.J. (1964). Ann. Med. Intern. Fenn. 53, 167.

LIPID COMPONENTS OF HUMAN MYELIN MEMBRANES FORMING THE MYELIN SHEATH

Hartmut Pilz

Neurologische Klinik der Universität Göttingen, (Germany).

Lipids and proteins form a specific membranous structure, the myelin sheath. Comparative analytical studies on grey and white matter and on myelination have suggested that there are certain typical myelin lipids, found in other organs at low concentration only (galactocerebrosides, sulfatides, sphingomyelins, unesterified cholesterol, phosphatidyl ethanolamine, phosphatidyl choline). Therefore it was assumed that in purified myelin material these lipids would be present in much larger proportions.

Specimens from several human brains following isolation and purification of myelin fractions using the technique of AUTILIO *et al*. (1964) were analysed by quantitative thin-layer chromatography (according to JATZKEWITZ, 1964). Our results show that the myelin lipids were not appreciably concentrated by this procedure in view of the high degree of purification of myelin. These methods are suitable for determining, not only the main lipid fractions, but also for differentiating the sphingolipids according to their fatty acid patterns. The investigations have shown furthermore, that also in human myelin there are more cerebrosides linked with hydroxylated fatty acids than cerebrosides with unsubstituted fatty acids. On the other hand, sulfatides linked with unsubstituted fatty acids predominate over sulfatides with hydroxylated fatty acids. Finally, more sphingomyelins containing fatty acids with a high carbon number (mainly around C_{24}) are found than those with shorter fatty acids (mainly around C_{18}). In addition to quantitative data on glycolipids and phospholipids of human myelin in relation to dry weight and total lipids, molar ratios will be presented.

The data suggest that lipids hitherto thought typical of myelin are also present in other fractions of white matter, related to myelin structurally or in other ways.

AUTILIO, L.A., NORTON, W.T. & TERRY, R.D. (1964). *J. Neurochem.*, *11*, 17. JATZKEWITZ, H. (1964). *Hoppe-Seylers Z. physiol. Chem.*, *336*, 25.

EFFECT OF AN ECTOSYLVIAN EPILEPTOGENIC FOCUS ON OXIDATIVE PHOSPHORYLATION IN THE CAT NEOCORTEX

C. Pintilie and Nella Mison-Crighel

Neurological Institute of the R.S.R. Academy, Bucharest

Experiments were performed on unanaesthetized cats in which an epileptogenic focus had been produced in the left ectosylvian gyrus by topical application of mescaline hydrochloride. The appearance and development of the epileptogenic focus were recorded by electroencephalograph.

Oxidative phosphorylation with succinate as substrate (using hexokinase and glucose as trapping agent for the ATP formed) was determined in homogenates of samples taken from the ectosylvian and marginal gyri bilaterally at different times after application of mescaline: 2 min for the first group of cats, 10 min for the second and 30 min for the third.

The P/O ratio was reduced in comparison with controls in all the neocortical areas studied at 10 and 30 minutes after topical application of mescaline, but the change was significant only in both marginal gyri. The low phosphorylation efficiency was due to a decreased phosphate uptake, while oxygen uptake was normal.

The present findings suggest that a disturbed oxidative phosphorylation is associated with changes in the excitability of the neocortex induced by a mescaline focus.

COMPARATIVE STUDIES OF FATTY ACID COMPOSITION OF PHOSPHOLIPIDS IN SUBCELLULAR FRACTIONS OF RABBIT BRAIN

L.F. Pomazanskaya, N.I. Pravdina and S.A. Zabelinsky

Institute of Evolutionary Physiology and Biochemistry of the Academy of Sciences of the USSR, Leningrad

The fatty acid composition of lecithins, ethanolamine-phosphatides, serinephosphatides and sphingomyelins was studied by gas-liquid chromatography in homogenates and in mitochondrial and microsomal fractions of the grey matter and myelin of the white matter of the rabbit brain.

Significant differences were demonstrated between individual phospholipids in the content both of saturated and of unsaturated fatty acids. The differences between the various glycerophosphatides from the grey matter were due mainly to quantitative variations in the relative proportions of different fatty acids. The general total composition of fatty acids was essentially similar. Lecithin was the most saturated of the glycerophosphatides. High contents of polyenic C_{20-22} fatty acids were typical of ethanolamine-phosphatides and serinephosphatides. The fatty acid composition of the sphingomyelin from the grey matter differed both quantitatively and qualitatively from that of the glycerophosphatides. Sphingomyelin contained fatty acids with chain length more than 22 C. Almost all the sphingomyelin fatty acids within the range 14-25 C were present as the saturated and corresponding mono-unsaturated acids.

The fatty acid composition of mitochondrial phospholipids was essentially similar to that of the microsomal fraction.

Myelin phospholipids differed significantly in their fatty acid composition from the corresponding phospholipids of the grey matter. The glycerophosphatides (even ethanolamine- and serinephosphatides) contained few polyenic fatty acids. On the contrary, the content of monoenic acids in these phospholipids was rather high.

MECHANISMS OF BIOSYNTHESIS OF PHOSPHATIDYLETHANOLAMINE AND PHOSPHATIDYLSERINE IN BRAIN TISSUE AT THE MICROSOMAL LEVEL. THE ROLE OF L-SERINE ETHANOLAMINE AND L-THREONINE ETHANOLAMINE PHOSPHODIESTERS

G. Porcellati

Istituto di Chimica Biologica,
Università di Pavia, Pavia.

Two phosphodiesters, L-serine ethanolamine phosphate (L-SEP) and L-threonine ethanolamine phosphate (L-TEP), have been crystallized from nervous tissues of many animal species, and their structure verified by comparison with the physical and chemical properties of the synthetic compounds (PORCELLATI _et al._, 1965). Recently, L-SEP and L-TEP have been thought to play a physiological role in the mechanism of brain phospholipid biosynthesis, acting possibly as natural precursors of the phosphatidylethanolamine, phosphatidylserine and presumably phosphatidylthreonine moieties, at a microsomal level (PORCELLATI, 1964).

The occurrence in brain microsomes of fish, amphibian and reptile species of an enzyme system which converts serine-labelled-L-[^{14}C]SEP to phosphatidylserine is now reported. The phosphatidylserine was identified by thin-layer procedures, by chromatography on silicic acid columns, by further incubation with a liver phosphatidylserine decarboxylase and by analysis of the products obtained after mild hydrolysis procedures. Results are also reported which show the participation of the other portion of the SEP molecule, namely of the phosphorylethanolamine molecule of ethanolamine-labelled-L-[^{14}C]SEP, in phosphatidylethanolamine biosynthesis by a similar microsomal preparation _in vitro_. The phosphatidylethanolamine was almost conclusively identified by comparison of the radioactive material with an authentic reference sample and by analysis of the products obtained after mild hydrolysis procedures.

Double label experiments, carried out with L-[^{3}H]serine [1,2-^{14}C]ethanolamine phosphate, incubated with brain microsomal fractions _in vitro_, have also indicated that the microsomal phosphodiesterase activity, which splits L-SEP to L-serine and phosphorylethanolamine, prior to conversion to lipid material, may possibly act as a metabolic regulator of the SEP-dependent-phospholipid biosynthesis in the microsomes.

Other results show that radioactivity is clearly incorporated into phospholipid on incubation of brain microsomes with threonine-labelled-L-[^{14}C]TEP, but into some as yet unidentified radioactive product(s).

Altogether the findings appear to indicate a physiological role for L-SEP (and probably for L-TEP), which can contribute both the moieties of L-serine and phosphorylethanolamine to the coupled biosynthesis of phosphatidylserine and phosphatidylethanolamine, thus favouring the slow, gradual, but regular deposition of lipid at the microsomal level.

PORCELLATI, G. (1964). Biochim. biophys. Acta, 90, 183.
PORCELLATI, G. et al. (1965). Comp. Biochem. Physiol., 14, 413.

CEREBRAL AMINO ACID METABOLISM IN BRAIN TUMOURS

G. Quadbeck

Heidelberg

The source of energy for the central nervous system is normally the oxidative metabolism of glucose. It could be demonstrated in patients suffering from pathological conditions that the uptake of oxygen was higher or lower than it should have been by calculation from the glucose uptake and the lactate, pyruvate and CO_2 output from the brain. In patients with brain tumours we measured with Dr. Hoyer the arterio-venous difference of 20 amino acids in the cerebral blood, using the method of MOORE and STEIN. We found that in various pathological conditions the brain can take up amino acids from the blood or deliver amino acids into the blood in considerable quantities. If there were a considerable amino acid uptake before removal of the tumour at operation, the uptake decreased greatly after operation. Similarly in patients where an output of amino acids into the venous blood from the brain was observed preoperatively, an uptake could be demonstrated after operation. In most cases the arteriovenous difference in proline fluctuated greatly. Other amino acids with considerable arteriovenous differences were asparagine, serine, glutamic acid, glycine, alanine, valine, ornithine and histidine. Of all the brain tumours astrocytomas had the largest arteriovenous difference for amino acids. In normal brains the difference between the total amino acids in arterial and venous blood was nearly zero, because a small uptake of one or other amino acid by the brain was compensated by output of another amino acid. It may be concluded, that in man in both pathological and normal conditions there is a considerable exchange of amino acids between blood and brain tissue.

CATIONIC STIMULATION OF RNA SYNTHESIS IN RAT BRAIN CORTEX SLICES

J.H. Quastel, Carol Prives and Hansa Menon

Neurochemistry Section, Kinsmen Laboratories, University of British Columbia, Vancouver.

The rate of incorporation of labelled uridine into RNA in adult rat brain cortex slices incubated in a Ringer-phosphate-glucose medium at 37° for 30 minutes was increased by increase of KCl concentration to 15 mM although there was no increased rate of respiration. With a further increase of KCl concentration to 100 mM the rate of incorporation was markedly diminished, the rate of respiration being considerably enhanced under these conditions. The stimulation also occurred when labelled adenine or labelled orotic acid was used as a precursor. The rate of incorporation of labelled uridine into the acid soluble nucleotides was not affected by 15 mM-KCl. The presence of ouabain (10^{-6} M) inhibited incorporation of labelled uridine into RNA but had only little effect on the stimulated incorporation in presence of 15 mM-KCl. Possibly this was due to the well known antagonism between ouabain and K^+. There was no evidence of stimulation by 15 mM-KCl of the rate of incorporation of labelled uridine into RNA in the isolated brains of new-born rats although the inhibition of incorporation by 100 mM-KCl was apparent. The rate of incorporation was, however, considerably higher in new-born rat brain than in adult rat brain.

OLFACTION AND CHANGES IN BRAIN NUCLEAR RNA IN A TELEOST

D.A. Rappoport and H.F. Daginawala

Division of Molecular Biology, Department of Pediatrics,
The University of Texas Medical Branch, Galveston, Texas.

Studies were initiated to correlate biochemical changes in brain nuclear RNA induced by olfactory stimulation using sea water catfish (Galeichthys felis).

Salt water catfish, allowed to swim for one hour in sea water containing an odorant, morpholine (10^{-4} M), showed an increase in brain nuclear RNA and a change in base ratio in contrast to controls in plain sea water. These changes in brain nuclear RNA were reversed within 24 hours to the level of unstimulated controls when morpholine stimulated fish were transferred to fresh sea water.

Using a split-brain preparation in an isolated catfish head, one naris was lavaged with morpholine in sea water (10^{-6} M), while the other naris was lavaged with plain sea water. The stimulated half of the brain, compared to the unstimulated half, showed the same changes in nuclear RNA as those noted in free swimming catfish. Similar experiments using split-brain preparations, with amyl acetate as the odorant, also elicited an increase in brain nuclear RNA and a change in base ratio, but the change in base ratio was distinctly different from that induced by morpholine. Using camphor and menthol as odorants, respectively, the former induced an increase in brain nuclear RNA, while the latter caused a decrease, however, neither of these odorants changed the base ratio of the nuclear RNA. Since the odorants tested were novel to the catfish, tests were made using a natural food odorant. Shrimp extracts in sea water were lavaged through the catfish nares. The amount of brain nuclear RNA did not change appreciably in amount but the base ratio showed a definitive change. The presence of actinomycin did not affect the changes in brain nuclear RNA in catfish during olfactory stimulation.

Surface bipolar macroelectrodes were used to monitor the evoked EEGs from the olfactory tract and telencephalon during olfactory stimulation. A 40 cps burst was observed similar to those described as arising from cat brain under novel stimuli or on arousal. The relation of afferent impulses arising from olfaction, and the enzyme involved in brain nuclear RNA synthesis, are discussed.

^{14}C-LABELLING OF N-ACETYL-ASPARTIC ACID AND N-ACETYL-ASPARTYL-GLUTAMIC ACID IN MOUSE BRAIN TISSUE SLICES

K.L. Reichelt and E. Kvamme

Department of Neurochemistry,
The University Psychiatric Clinic, Vinderen, Oslo.

The labelling of NAA(N-acetyl-aspartic acid), NAAGA (N-acetyl-aspartyl-glutamic acid) and GSH (glutathione) was studied using [U-^{14}C]-labelled constituent amino acids, glutamine, glucose and [1-^{14}C]acetate. Mouse brain tissue slices (1 g) were incubated (37°, pH 7.4) in a medium (10 ml) containing (mM), phosphate, (10); Mg^{++}, (0.65); Ca^{++}, (1.3); K^{+}, (13) and the labelled compound, 0.01 mC. After deproteinization with perchloric acid in the cold and KOH neutralization, the extract (pH 7) was fractionated by column chromatography on Dowex 2 X-8, as described by CURATOLO et al (1965). Aliquots of each compound were lyophilized, hydrolyzed in 6 N-HCl, the constituent amino acids identified by two-way paper chromatography and ninhydrin treatment. Paper strips of each amino acid were counted (in toluene) using a Packard Liquid Scintillation Spectrophotometer, followed by elution and measurement of the ninhydrin colour. Acetate was determined as the hydroxamate derivative.

Following incubation for 60 minutes, labelling of NAA- and NAAGA-aspartate was found, and the specific activities (counts per μmole) varied from 1/2 to 1/7 of that of GSH-glutamate. Using the same isotopic compounds, the glutamate group of NAAGA was also labelled. The specific activity of NAAGA-glutamate followed the pattern of GSH-glutamate and free glutamate, but was considerably lower. In addition ^{14}C-incorporation into the acetate groups of NAA and NAAGA were found. However, the acetyl labelling seemed to follow a pattern different from that of the bound amino acids. The data indicate that NAA and NAAGA are not metabolically inactive as previously stated by TALLAN (1957).

CURATOLO, A., D'ARCANGELO, P., LINO, A. & BRANCATI, A. (1965). J. Neurochem., 12, 339. TALLAN, H.H. (1957). J. biol. Chem., 224, 41.

THE COMPARATIVE PRODUCTION OF γ-AMINOBUTYRATE IN BIDIRECTIONALLY BRED BEHAVIOURAL STRAINS

J.T. Rick and G.A. Kerkut

Department of Physiology and Biochemistry,
The University of Southampton, England.

The basic strategy of the experiments described was to employ animals which had been bidirectionally selected and inbred on some specific behavioural parameter - i.e. animals whose differential patterns of behaviour had been genetically "frozen" - and to investigate the intermediary metabolism of the brain in such animals for concomitant variation with behavioural selection. By these means it is assumed that one has at one's disposal animals whose behavioural differences reflect a functional variation in brain metabolism which can be demonstrated by routine biochemical analysis.

The intermediary metabolism of various brain areas of the Maudsley 'Reactive' and 'Nonreactive' strains was studied. These strains are descended from a heterogeneous group of albino rats selectively inbred for over twenty generations, the selection being based on high and low scores of defaecation in an 'open field' situation (BROADHURST, 1964). The rate of production of γ-aminobutyrate from [U-^{14}C]glucose in the cerebral hemispheres of the 'Nonreactive' strain was significantly higher than that observed in the 'Reactive' strain.

The results are interpreted in the context of other studies where γ-aminobutyrate activity has been related to behavioural measures.

BROADHURST, P.L. (1964). In Animal behaviour and drug action. Ciba Foundation Symposium. Edited by Hannah Steinberg, A.V.S. de Reuck & Julie Knight, p. 224. London: Churchill.

ISOLATION AND PROPERTIES OF ISOLATED NERVE ENDING MEMBRANES AND JUNCTIONAL COMPLEXES

E. De Robertis, G. Rodríguez de Lores Arnaiz,
M. Alberici, J.M. Azcurra and S. Fiszer

Instituto de Anatomia Genral y Embriologia,
Facultad de Medicina, Universidad de Buenos Aires.

The osmotic disruption of isolated nerve endings - first developed to separate the synaptic vesicles (DE ROBERTIS *et al*., 1962) - is used as a first step to isolate, in a sucrose gradient, various layers of nerve ending membranes (DE ROBERTIS *et al*., 1966). These membranes are in continuity with the synaptic structures which constitute the *junctional complexes*.

The various populations of nerve ending membranes were characterized by:

a) a group of membrane-bound enzymes i.e. acetylcholinesterase, Na^+-K^+-ATPase, K^+phosphatase, glutamine synthetase, adenyl cyclase and cyclic phosphodiesterase. The last two enzymes were studied in collaboration with Profs. Sutherland and Butcher.

b) the binding capacity of radioactive cholinergic blocking agents

c) some differences in chemical composition particularly in the content of gangliosides (LAPETINA *et al*., 1967).

Recently the treatment of the nerve ending membranes with Triton X-100 has lead to the concentration of the junctional complexes with dissolution of most of the nerve ending membranes. These retain the receptor properties but lose a large proportion of the membrane-bound enzymes.

These studies open the possibility of investigating at a macromolecular level the brain receptors.

Supported by Grants from the Consejo Nacional de Investigaciones Científicas y Técnicas NIH, Grant NB 06953-01

DE ROBERTIS *et al*. (1962). *Nature*, *194*, 794. DE ROBERTIS *et al*. (1966). *Life Sci*., *5*, 577. LAPETINA *et al*. (1967). *Biochim. biophys. Acta.* in press.

ASSOCIATION OF COMPONENTS OF THE γ-AMINOBUTYRIC ACID (GABA) SYSTEM WITH SYNAPTIC VESICLES

E. Roberts and K. Kuriyama

Department of Biochemistry,
City of Hope Medical Center, Duarte, California

Evidence is accumulating which suggests strongly that in vertebrates γ-aminobutyric acid (GABA) may be an inhibitory transmitter liberated presynaptically on excitation in the central nervous system of some neurons with an inhibitory function, exerting its action by increasing the Cl^- conductance of postsynaptic membranes. Details of the mechanisms of storage and release of GABA or of its mode of action on neural membranes at a molecular level still are not known. Our work now is directed to examination of the possibility that the association of the GABA system with synaptic vesicles (S-V) may be related to presynaptic formation and storage of GABA prior to its release. Morphologically homogeneous (mean diameter, 600 A°) S-V were isolated from mouse brain homogenate by differential centrifugation and hypo-osmotic treatment of nerve endings. GABA, glutamic acid, and glutamic decarboxylase activity (GAD) were found to be associated with the electron microscopically characterized S-V, but no GABA transaminase activity was found. Discontinuous density gradient fractionation of the vesicles gave four separate fractions of particles which appeared morphologically similar and which had approximately the same contents of the above constituents on a protein basis as the unfractionated S-V. Both the original and fractionated S-V possessed the ability to bind exogenously added [^{14}C]GABA at 0° C by a process which is not energy-dependent. Thus, there is a close association of the constituents of the GABA system with S-V. The effects of changing conditions and of drugs on this association will be discussed.

ENZYME HISTOCHEMICAL STUDIES ON THE CEREBRAL CORTEX OF THYROID HORMONE-DEFICIENT RATS

N. Robinson, J.G. Nievel and J.T. Eayrs

Institute of Neurology, The National Hospital, London, W.C.1. and Institute of Psychiatry, The Maudsley Hospital, London, S.E.5.

Changes in the behaviour of the thyroid-deficient developing animal and in the capacity for learning have been reported (BRADLEY, EAYRS & SCHMALBACH, 1960; BRADLEY, EAYRS, GLASS & HEATH, 1961; EAYRS & LEVINE, 1963). Studies on the cortical histology indicate changes in the proximity and structure of cortical nerve cell bodies (EAYRS & TAYLOR, 1951) and their processes (EAYRS, 1955; EAYRS & HORN, 1955), the capillary bed (EAYRS, 1954) and retardation of myelination (BARRNETT, 1943). Biochemical investigations show a moderate decrease in cerebral activity of two enzymes (HAMBURGH & FLEXNER, 1957).

The distribution and activity of 12 enzymes concerned in important metabolic pathways of the CNS were examined in the cerebral cortex of 30-day old rats thyroidectomized 2 days post-partum. A moderate increase in α-glycerophosphate dehydrogenase was apparent in blood vessels and capillaries. Succinate and glucose 6-phosphate dehydrogenases were decreased in nerve cell perikarya. Acid phosphatase was decreased in nerve cell perikarya and increased in glia. Alkaline phosphatase and Ca^{2+}-ATPase were decreased in blood vessels and capillaries. Nerve fibres, blood vessels and capillaries showed a raised 5-nucleotidase whilst monoamine oxidase was decreased in nerve cell perikarya and increased in nerve fibres. $NADH_2$-diaphorase, Mg^{2+}-ATPase, thiamine pyrophosphatase and acetylcholinesterase were normal.

Phosphatase decreases in the blood vessels suggested some alteration in the control of the permeability and supply of free energy at the vascular walls (HESS & POPE, 1959). The lower MAO activity in nerve fibres, normally prominently stained at 30 days, reflected histological observations on retardation of myelination. The changes in acid phosphatase activity in the absence of oxidative enzyme elevations in glial cells did not parallel findings in these cells in other pathological conditions. The several enzymes exhibiting a lower activity in nerve cell perikarya reflected some alterations in metabolism, but no one enzyme showed a gross abnormality in activity suggestive of a principal factor in a derangement in the control of the metabolism in cortical nerve cells.

BARRNETT, R.J. (1948). In The hormones: physiology, chemistry and applications. Edited by G. Pincus & K.V. Thimann, Vol.2 p.235. New York, Academic Press. BRADLEY, P.B., EAYRS, J.T., GLASS, A. & HEATH, R.W. (1961). Electroenceph. Clin. Neurophysiol. 13, 557. BRADLEY, P.B., EAYRS, J.T. & SCHMALBACH, K. (1960). Electroenceph. Clin. Neurophysiol., 12, 467.
EAYRS, J.T. (1954). J. Anat. Lond., 88, 164. EAYRS, J.T. (1955). Acta Anat., 25, 160.

EAYRS, J.T. & HORN, G. (1955). Anat. Rec. 121, 53. EAYRS, J.T. & LEVINE, S. (1963). J. Endocr., 25, 505. EAYRS, J.T. & TAYLOR, S.H. (1951). J. Anat. Lond., 85, 350. HAMBURGH, M. & FLEXNER, L.B. (1957). J. Neurochem., 1, 279. HESS, H.H. & POPE, A. (1959). J. Neurochem., 3, 287.

METABOLISM OF ISOLATED NEURONAL AND GLIAL CELLS

S.P.R. Rose

Medical Research Council Metabolic Reactions Research Unit, Imperial College of Science, London, S.W.7.

Further aspects of hexose and amino acid metabolism of neuronal and glial cells isolated by the method of ROSE (1965; 1967; see also BRADFORD & ROSE, 1967) have now been examined. Cells isolated by this technique were compared with those prepared by the methods of RAPPAPORT (1966) and SATAKE & ABE (1966). Compared to cells prepared by these methods, oxygen consumption with glucose was 10-fold higher, CO_2 production 6-fold higher, lactate production 17-fold higher and incorporation of $[H^3]$lysine into protein 2.5-fold higher.

Cells isolated by the method of ROSE (1965) maintain a high internal concentration of free amino acids; the concentration in the neuronal fraction of aspartate and glutamine was 2-fold, of alanine 3-fold, and of GABA one-half that in the glial fraction. The metabolic fate of glutamate in the neuronal fraction was different from that in the glial fraction, there being relatively less incorporation of label from $[U\text{-}C^{14}]$glutamate into CO_2 and citric acid cycle intermediates and more into acid-insoluble material. Incorporation of other amino acids, such as $[H^3]$lysine, into protein was also up to 2-fold higher in the neuronal than the glial fraction, though still less than in the intact slice.

These studies, and their implications for the validity of the use of isolated cell preparations, will be discussed.

BRADFORD, H.F. & ROSE, S.P.R. (1967). J. Neurochem. In press. RAPPAPORT, C. (1966). Proc. Soc. Exp. Biol. Med., 121, 1010. ROSE, S.P.R. (1965). Nature, Lond., 206, 621. ROSE, S.P.R. (1967). Biochem. J., 102, 33. SATAKE, M. & ABE, S. (1966). J. Biochem. (Tokyo), 59, 72.

THE OCCURRENCE AND METABOLISM OF TRIGLYCERIDES IN BRAIN

C.E. Rowe

Department of Physiological Chemistry,
University of Birmingham, Birmingham, 15.

A procedure has been devised for the isolation and estimation of triglyceride in brain. This is based on thin layer chromatography (SKIPSKI *et al.*, 1965) followed by measurement of the infra-red absorption, in the range 1740-1745 cm.$^{-1}$, of the C=O group (FREEMAN *et al.*, 1957). Triglyceride was isolated from whole mouse brain, whole rat brain, rat brain grey and white matter, and incubated slices of rat brain grey matter, in yields of 0.1-0.3 μmoles/g of fresh tissue.

The predominant fatty acid esters obtained from the methanolysis of triglyceride were usually palmitate, stearate and oleate, together with small quantities of esters with the chromatographic properties of unsaturated acids containing odd numbers of carbon atoms. When present, these latter acids were also present in unesterified fatty acids isolated from the same tissue.

[1-^{14}C]Acetate was rapidly incorporated into the fatty acid esters in triglyceride of rat brain grey matter slices incubated for 1 hr. at 37° in Krebs-Ringer bicarbonate containing glucose (1.1 x 10^{-2} M) and E.D.T.A. (3 x 10^{-5} M; disodium salt). The incorporation of acetate per mole of triglyceride was similar to the incorporation per mole of unesterified fatty acid. Radioactive triglyceride was recovered from both mitochondrial and microsomal fractions prepared from incubated slices.

When incubated with pancreatic lipase, radioactive triglyceride, prepared from incubation of slices with [1-^{14}C]acetate, yielded radioactive monoglyceride and diglyceride in addition to radioactive fatty acid. Since pancreatic lipase produces β-monoglyceride (SAVARY & DESNUELLE, 1956, 1955; MATTSON & BECK, 1956, 1955) both α and β fatty acids of the triglycerides must have been labelled with ^{14}C.

In slices incubated in buffer containing 3 x 10^{-4} E.D.T.A. the concentration of endogenous triglyceride decreased at a rate of 0.07 μmoles/hour/g of tissue.

The author is grateful to the Medical Research Council of Great Britain for a grant for expenses in aid of this work.

FREEMAN, N.K., LINDGREN, F.T., YOOK, C.NG, & NICHOLS, A.V. (1957). J. biol. Chem., 227, 449. MATTSON, F.H. & BECK, L.W. (1955). J. biol. Chem., 214, 115. MATTSON, F.H. & BECK, L.W. (1956). J. biol. Chem., 219, 735. SAVARY, P. & DESNUELLE, P. (1955). Compt. rend., 240, 2517. SAVARY, P. & DESNUELLE, P. (1956). Biochim. biophys. Acta, 21, 349. SKIPSKI, V.P., SMOLOWE, A.F., SULLIVAN, R.C. & BARCLAY, M. (1965). Biochim. biophys. Acta, 106, 386.

LIPID CHARGE AND MEMBRANE STRUCTURE

S. Samuels and F.P. Aleu

New York University Medical Center, New York

Membrane cytoplasmic aggregates have been observed by electron microscopy in the different lipidoses and in Alzheimer's Disease. Our previous research (SAMUELS *et al.*, 1965) suggested that these structures arise by the spontaneous interaction of the component molecules in response to the localized excess of one or more particular lipids. Structural variants of the aggregates have been observed within the affected cells and, as a result of varying the conditions in our *in vitro* studies. The present report is a further study of the effect of the lipid and inorganic constituents on the structure of membranous arrays produced *in vitro* and a comparison with the isolated membranous cytoplasmic bodies (MCB) of Tay Sachs disease.

It has been found that a mobile cation is essential for maintaining the structural integrity of the Tay Sachs MCB and for permitting the aggregation of the organic components to form MCB-like structures *in vitro*. Organic, as well as inorganic, cations can serve in this role, but calcium is more effective than either sodium or potassium, and they in turn much more so than glycine in protecting the structural integrity of isolated MCB. The ultrastructure of phospholipid aggregates appears to be related to the net charge on the esterified phosphate moiety. This may be related to the molecular structure of natural membranes and may explain in part, the reason for the variety of lipid types in natural membrane structures.

SAMUELS *et al.* (1965). *J. Neuropath. exp. Neurol.*, *24*, 256.

IONIC COMPOSITION OF BRAIN SLICES ISOLATED FROM THE RAT

E. Schoffeniels*

Department of Biochemistry, University of Liège

The electrophysiological data obtained by measuring the resting potentials of neurons are generally interpreted as indicating a high potassium content (ca 138 μE/ml cell water) and a low sodium content (ca 35 μE/ml cell water). Chemical analysis of brain slices prior to any contact with artificial saline leads to the same result, if one postulates that 5 to 10% of the fresh weight is extracellular water.

When a brain slice is incubated in saline, there is a rapid shift in the ionic composition. The values reported in the literature are variable, depending not only on the species used but also on the investigator. While one must interpret the results of intracellular analysis with caution, a reproducible method should give valuable information concerning the metabolic state of isolated brain cells.

We have devised a standard experimental procedure. A new technique of slicing the brain has been set up, making use of a nylon thread to cut (within 1 sec) slices of 0.4 to 0.5 mm thick 3 to 4 min after killing the animal. Our study deals with the changes in time in fresh weight, dry weight, inulin space and sodium and potassium content of slices of rat brain before and after contact with an artificial saline.

The results obtained depended on whether the slice was incubated in a small volume of liquid (15 ml) or in a stream of saline (7,5 L/h). They show that within 5 min a new cationic equilibrium is reached in the cell. It is characterized by an increase in the intracellular sodium (from 54 μE/ml to 110 μE/ml) and a decrease in the intracellular potassium (from 143 μE/ml to 70 μE/ml.

The meaning of the inulin space, the uptake of water, the loss of dry weight and the effect of a crude brain extract on the ionic composition of the slice will be discussed.

*The results presented have been obtained in collaboration with Drs. G. Franck and M. Cornette.

THE EFFECT OF LITHIUM ON CEREBRAL MONOAMINE NEURONS

Mogens Schou, Hans Corrodi, Kjell Fuxe and Tomas Hökfelt

State Mental Hospital, Risskov, Denmark.

Lithium is not only thereapeutically active against manic phases of the manic-depressive psychosis, but according to recent observations (BAASTRUP & SCHOU, 1967) it is also able to prevent further psychotic attacks when given continuously to patients with recurrent manic-depressive psychoses and recurrent endogenous depression. Since anti-depressant drugs have been shown to interfere with the metabolism and transport of cerebral monoamines, a study was made of the effect of lithium on the monoamine neuron systems. Lithium was administered intraperitoneally to rats in varying doses, and the animals were killed after varying periods. A potent inhibitor of amine synthesis was also administered to the animals in some experiments. Serum and brain lithium concentrations were determined flame-photometrically, and the levels of noradrenaline, dopamine and serotonin in the brains were determined by histochemical as well as by chemical methods. Results of these studies will be reported and discussed.

BAASTRUP, P.C. & SCHOU, M. (1967). Arch. gen. Psychiat., 16, 162.

INCORPORATION OF EXOGENOUS ACETYLCHOLINE, CHOLINE AND ATROPINE IN MOUSE BRAIN CORTEX SLICES

J. Schuberth and A. Sundwall

Research Institute of National Defence, Sundbyberg 4, Sweden.

It has recently been shown that choline (Ch) and acetylcholine (ACh) are taken up against a concentration gradient when brain cortex slices are incubated in a suitable medium (SCHUBERTH et al., 1966; POLAK & MEEUWS, 1966). Using tritium-labelled atropine we have shown that this drug is also accumulated in cortex slices against a concentration gradient. In the present communication we have compared the uptake of Ch, ACh and atropine with respect to inhibition by drugs and subcellular distribution.

Hemicholinium (HC-3), eserine, atropine, oxotremorine and morphine were found to be competitive inhibitors of the uptake of ACh with K_i values 0.5×10^{-5}, 0.7×10^{-5}, 1.6×10^{-5}, 2.3×10^{-5} and 3.6×10^{-5} M. Of these drugs only HC-3 was found to inhibit Ch uptake (K_i 0.6×10^{-4} M). Thus HC-3 is a ten times stronger inhibitor of ACh transport than of Ch transport. Atropine uptake was not affected by any drug tested.

Studies on the subcellular localization of the drugs taken up by the slices revealed that the main part of the radioactivity was concentrated on the nerve-ending fraction following incubation with Ch or ACh. In contrast, no specific subcellular localization of radioactivity was found after incubation with atropine. In the latter case the relative distribution of radioactivity between the different fractions coincided with that following incubation with tritiated water.

The results thus clearly show that Ch and ACh uptake differ from that of atropine, both with respect to susceptibility to drugs and with respect to subcellular distribution.

POLAK, R.L. & MEEUWS, M.M. (1966). Biochem. Pharm., 15, 989.
SCHUBERTH, J., SUNDWALL, A., SORBO, B. & LINDELL, J.O. (1966). J. Neurochem., 13, 347.

THE IDENTIFICATION AND MEASUREMENT OF MYO-INOSOSE-2 AND SCYLLO-INOSITOL IN MAMMALIAN TISSUES

W.R. Sherman and M.A. Stewart

Washington University School of Medicine
St. Louis, Missouri

POSTERNAK et al. (1963) have shown that scyllo- and myo-inositol are interconverted by the whole rat and they provide evidence that myo-inosose-2 is an intermediate in this transformation. Myo-inosose-2 has, however, not previously been observed to occur naturally. Further, scyllo-inositol has not previously been localized in mammalian tissues, though it has been found in whole animals and in mammalian urine.

Tissue samples were extracted with water at 100° and the supernatant deproteinized and lyophilized. The dried samples were treated with pyridine: hexamethyldisilazane: trimethylchlorosilane 17:2:1 and the volatile trimethylsilyl carbohydrate ethers thus obtained were subjected to gas chromatography.

Scyllo-inositol was detected and measured by hydrogen flame detector. Chromatography on different columns assured homogeneity of the eluted sample. Trimethylsilyl scyllo-inositol was isolated by gas chromatography from dog sciatic nerve and from rabbit kidney. In both instances the material obtained was identical with authentic trimethylsilyl scyllo-inositol by infra-red spectroscopy.

Myo-inosose-2 was detected and measured by a tritium-foil electron capture detector. Gas chromatographic collection from rabbit nerve gave approximately 75 nanograms of trimethylsilyl myo-inosose-2 which was identified by 1) retention time comparison with authentic inosose on different columns 2) conversion to the oxime and 3) reduction by sodium amalgam to myo-inositol. The latter were identified by gas chromatographic comparison with authentic materials. Work is underway to obtain sufficient material for spectroscopic identification of myo-inosose-2 from calf brain.

Typical levels of these inositols which have been measured chromatographically are as follows: Scyllo-inositol (millimoles/kg wet) : Rabbit sciatic nerve 0.13. Rat sciatic nerve 0.22; spinal cord 0.18; brain 0.16; kidney 0.45; serum <0.001. Myo-inosose-2 (micromoles/kg wet) : Rat brain 1.6; sciatic nerve 0.6; kidney 3.2. Rabbit cerebellum 1.3; grey matter 0.9; white matter 0.6; sciatic nerve 0.4; optic nerve <0.2; testis 3.3; kidney 1.9; liver 0.7; spleen 0.6; epididymis 0.3; heart 0.2; muscle, lens, lung ≤0.2; pancreas 0.

POSTERNAK et al. (1963). Helv. chim. Acta, 46, 2676.

RELATIONS BETWEEN THE TISSUE ACID-BASE AND THE TISSUE ENERGY METABOLISM IN THE BRAIN

B.K. Siesjö and L. Granholm

Department of Neurosurgery, University of Lund and the Neurosurgical Service A, University Hospital, Lund, Sweden.

There are several reasons to assume that important relations exist between the acid-base metabolism and the energy metabolism of brain tissue. Thus, phosphate compounds probably form some of the most important intracellular buffers with the phosphate group of phosphocreatine acting as a potential hydrogen ion acceptor, and ATP possibly as an energy source for active transport of H^+ out of the cells. Moreover, lactic acid formed during tissue hypoxia should drastically influence intra- and extracellular hydrogen ion activities. Finally, evaluation of acid-base changes is imperative if the redox state of the $NAD^+/NADH$ system should be evaluated from concentration relations in the coupled lactate/pyruvate system.

In order to acquire the basic information necessary for the evaluation of such interrelations in tissue hypoxia and tissue asphyxia, experiments were made in cats and rats with measurements of lactate and pyruvate in CSF and arterial blood, as well as of lactate, pyruvate, ATP, ADP, AMP, phosphocreatine and creatine in brain tissue in various systemic acid-base shifts. These include both respiratory and non-respiratory acidosis and alkalosis of acute and chronic duration. In all these experiments the tissue was frozen _in situ_ and the metabolites were analysed by enzymic methods. The results will be discussed in terms of the influence of glycolysis and of the metabolism of organic phosphates on the regulation of intra- and extracellular hydrogen ion activities in the brain.

IN VITRO SYNTHESIS OF MYELIN IN NORMAL AND DEMYELINATING RATS

Marion Edmonds Smith

Neurology Service, Veterans Administration Hospital and Department of Medicine (Neurology), Stanford University School of Medicine, Palo Alto, California

A method measuring *in vitro* synthesis of myelin was used for determining its metabolic state during demyelination. One spinal cord from a rat 40-50 days of age was sliced in half, and incubated two hours in [U-^{14}C]glucose. The slices were washed, homogenized and used for preparation of purified myelin by high speed centrifugation in a discontinuous sucrose gradient. When the lipids were extracted from the purified myelin and separated by thin layer chromatography, inositol phosphatide and lecithin showed the highest rate of uptake of ^{14}C from glucose. Serine phosphatide, ethanolamine phosphatide, sphingomyelin, cerebroside, cerebroside sulphate and cholesterol showed lower rates of uptake. This pattern of uptake is in agreement with previous *in vivo* experiments relating to rates of turnover of myelin lipids. The non-myelin fraction incorporated 6-15 times as much ^{14}C/mg lipid as the myelin.

Experimental allergic encephalomyelitis (EAE) was produced by injection of whole guinea-pig spinal cord in Freund's Adjuvant into the hind-foot pad of the rat. The uptake of [^{14}C]glucose into spinal cord myelin was measured in different stages of EAE. In the earlier stages (preparalytic) the uptake into myelin lipids was depressed, but with the progression of the disease through paralysis, the *in vitro* labelling of myelin lipid was increased above the normal. The changes in rate of uptake of ^{14}C into each of the individual myelin lipids was in proportion to the changes in the total lipids. These changes in myelin metabolism may be correlated with other abnormalities found in nervous tissue in EAE (SMITH, 1966), and the results indicate that paralysis resulting from demyelination is associated with an abnormal metabolism both inside and outside the myelin sheath.

SMITH, M.E. (1966). Nature, Lond., 209, 1031.

ACTIVATION OF SINGLE NEURONS IN RAT BRAIN BY L-GLUTAMIC ACID BEFORE AND AFTER INHIBITION OF THE DECARBOXYLASE: INTERACTIONS OF PYRIDOXAL-5'-PHOSPHATE

F.A. Steiner and K. Ruf

Department of Experimental Medicine, F. Hoffmann-La Roche Ltd. Basel and Institute for Brain Research, University of Zurich.

Minute amounts of L-glutamic acid administered into the immediate extracellular environment by microelectrophoresis (under circumvention of the blood brain barrier) excite single neurons in the rat brain. This cellular excitation is enhanced by systemic administration of hydrazides (e.g. thiosemicarbazide) a known inhibitor of vitamin B_6 dependent enzymes such as L-glutamic acid decarboxylase. This inhibition of L-glutamic acid decarboxylase is partially antagonized by concomitant microelectrophoretic administration of the enzymic cofactor, pyridoxal-5'-phosphate. These changes, previously described in biochemical and clinical terms, were thus traced to the level of the single cell.

GLYCOLIPIDS OF PERIPHERAL NERVE IN METACHROMATIC LEUCODYSTROPHY (MLD)

Pierre Stoffyn and Michael Malone

McLean Hospital, Research Laboratories, Belmont, Mass., and Department of Biological Chemistry, Harvard Medical School, Boston, Mass.

Previous studies have shown that the chemical structure of sulphatides, which are abnormally accumulated in brain white matter and kidney of patients with MLD, remains normal with the sulphate group at carbon 3 of galactose. Moreover, the fatty acid composition of the sulphatides, which is characteristically different in each of those normal organs, is practically unchanged in the disease. These studies have now been extended to the sulphatides and cerebrosides isolated from peripheral nerve of normal subjects and patients with MLD. Methylation of the sulphatides, followed by methanolysis and hydrolysis of the carbohydrate fraction, resulted in the isolation and characterization of 2,4,6-tri-O-methyl-D-galactose. This shows that the sulphate group in sulphatides of MLD peripheral nerve is located at carbon 3 of galactose as in brain and kidney. In MLD peripheral nerve, the sulphatides are increased about two-fold in concentration and, as in normal nerve, they contain mainly hydroxylated fatty acids. The fatty acid distribution, analysed by GLC, does not differ significantly from the pattern ovserved in sulphatides from normal nerve. These, like the cerebrosides of this tissue, are characterized by a larger proportion of short chain fatty acids (C_{12}-C_{18}) than the glycolipids of brain white matter and kidney.

In conclusion, the retention of a normal chemical structure and of a normal, organ-specific, composition in fatty acids indicates that the metabolic error leading to the pathological accumulation of sulphatides in different organs in MLD is not attributable to the biosynthesis of an abnormal sulphatide molecule.

Supported by United States Public Health Service Grants NB 06370 and NB 02840 and a Clinical Investigation by the Veterans Administration.

ISOLATION, CHEMICAL CHARACTERIZATION AND COMPARISON OF THE MEMBRANOUS CYTOPLASMIC BODIES FROM BRAINS OF TAY-SACHS DISEASE AND G_{M1}-GANGLIOSIDOSIS (GENERALIZED GANGLIOSIDOSIS)

Kunihiko Suzuki, Kinuko Suzuki and Gloria C. Chen

The Saul R. Korey Department of Neurology and Department of Pathology (Neuropathology), Albert Einstein College of Medicine, New York, N.Y.

SAMUELS et al. (1963) isolated and analysed the membranous cytoplasmic bodies (MCB) from a brain with Tay-Sachs disease. Neurons of the newly recognized G_{M1}-gangliosidosis (generalized gangliosidosis, systemic late infantile lipidosis) contain morphologically similar abnormal organelles. We have isolated MCB fractions from a Tay-Sachs brain and two G_{M1}-gangliosidosis brains. This is the first report of the chemical composition of MCB from G_{M1}-gangliosidosis. Also, recent improved analytical techniques have enabled us to obtain some new chemical findings on Tay-Sachs MCB not reported previously by SAMUELS et al. Under electron microscope, the morphology of the lamellar structures was fairly well-preserved with virtually no contaminants detected. The range of the chloroform-methanol insoluble residue was 6 to 9% of dry weight. About twice as much protein was initially soluble into chloroform-methanol and later rendered insoluble by repeated drying of the extract. This "proteolipid protein" contained significantly higher proportions of neutral amino acids as compared to the insoluble protein. Gangliosides constituted 30 to 40% of dry weight. They consisted of almost pure single molecular species in each disease, Tay-Sachs ganglioside (G_5, G_{M2}) in Tay-Sachs MCB, and the normal major monosialoganglioside (G_4, G_{M1}) in G_{M1}-gangliosidosis. Series of ceramide hexosides were present. Ceramide trihexoside was predominant in Tay-Sachs MCB, and ceramide tetrahexoside in G_{M1}-gangliosidosis MCB. Ceramide dihexoside was characterized as lactoside. Both sugar chromatography and borate-impregnated silica gel TLC indicated that most of cerebroside was glucocerebroside, although galactocerebroside was detectable. Cholesterol constituted 40 to 50% of total lower phase lipids. The molar ratios of phosphatide:cholesterol:ganglioside were 1.0:2.6:1.2 for Tay-Sachs MCB, and 1.0:2.9:1.1 and 1.0:2.8:1.1 for each of the MCB fractions from G_{M1}-gangliosidosis. Therefore, these abnormal membranous structures appear to have a very similar chemical composition except for the kinds of gangliosides and ceramide hexosides, which are distinct and characteristic in the respective diseases.

SAMUELS et al. (1963). J. Neuropath. exp. Neurol., 22, 81.

FREE AND PROTEIN-BOUND LYSINE FLUX IN DEVELOPING RABBIT BRAIN

Kenneth F. Swaiman and Richard N. Wolfe

Department of Psychiatry and Neurology, Medical School, University of Minnesota, Minneapolis, Minn.

There is some evidence that there is increased flux of lysine from plasma to the non-protein nitrogen pool of brain and increased flux of lysine from the free amino acid pool to protein in immature animals over mature animals. Studies of cerebral amino acid flux at different ages have been carried out systematically during the "critical period" of brain development in the rabbit.

Rabbits 1, 4, 8 and 12 days of age were injected intraperitoneally with L-[U-^{14}C]lysine (8.0 μcuries per 100 g body weight in a volume of 0.4 ml per 100 g body weight). The animals were decapitated at 12, 20, 40 and 60 minutes after injection. Ion-exchange resin studies of amino acid concentration and liquid scintillation counting of lysine radioactivity were carried out. The data consisting of specific activities of lysine at different elapsed times after injection in plasma, free amino acid fraction and protein-bound amino acid fraction were employed to determine flux of lysine in the immature rabbit brain.

It is noted that in excess of 30% of the radioactivity is present in amino acids other than lysine in the free amino acid fraction by 60 minutes after injection in these immature animals. Radioactivity in the plasma has fallen by 10 minutes after injection and falls to less than 10% of the 10 minute activity by 60 minutes in all ages studied.

The pattern of flux of lysine from plasma to the free lysine pool reaches its peak at about 20 minutes in the animals less than 8 days old but does not reach its peak until later in the 12 day old rabbit brain. Flux of lysine into protein during the 60 minutes of the experiment is relatively constant but is slightly more rapid in the less mature animals.

The data suggest that there is a difference in the flux of lysine from plasma into the free amino acid pool between the younger animals and the 12 day old animals. This experiment did not conclusively demonstrate a difference of lysine flux into protein between the various ages studied.

EFFECTS OF OUABAIN ON ISOLATED GUINEA-PIG CEREBRAL CORTEX

Phillip D. Swanson

Division of Neurology
University of Washington School of Medicine, Seattle 98105

The cardiac glycoside ouabain profoundly alters respiration and electrolyte content of cerebral cortex slices. Tissue content of Na^+ increases and that of K^+ decreases. However, the effect on respiration is variable. When Ca^{++} is present in the medium, respiration increases while respiration decreases if Ca^{++} is omitted. Since ouabain inhibits the cation-dependent ATPase, it has been suggested that the fall of respiration in Ca^{++}-free media results from decreased utilization of high-energy phosphate for cation transport.

In the present experiments with guinea-pig cerebral cortex slices, ouabain in concentrations of 10^{-5} to 10^{-4} M brought about expected shifts of Na^+ and K^+ contents. Ouabain also caused marked depression of slice content of creatine phosphate both in the presence or absence of Ca^{++}. In the absence of Ca^{++}, the rate of fall of creatine phosphate was somewhat slower than that of K^+. In the presence of Ca^{++}, slices lost creatine phosphate and K^+ at about the same rate.

Slices incubated in the presence or absence of 5×10^{-6} M-ouabain lost K^+ and gained Na^+ in response to electrical stimulation. However, ouabain-treated slices failed to utilize creatine phosphate during 10 min. of electrical stimulation. Ten minutes after electrical stimulation, in contrast to control slices, ouabain-treated slices had failed to recover K^+ or to lose Na^+. At this time, creatine phosphate contents of ouabain-treated slices had fallen from 0.9 to 0.5 μmoles/g, while control levels rose from 0.3 to 1.0 μmoles/g.

Ouabain did not affect oxidative phosphorylation or respiratory control ratios when added directly to mitochondria prepared from beef or guinea-pig brain.

The results suggest that ouabain does interfere with active transport of Na^+ and K^+ in cerebral tissues. However, the decrease in respiration which occurs in Ca^{++}-free media is probably due to interference with energy-producing mechanisms which follow the marked K^+ loss from the slice and is not, therefore, a good indication of the proportion of tissue metabolism used for active cation transport.

REGULATION BY PYRUVATE KINASE OF AEROBIC GLYCOLYSIS IN BRAIN TISSUE *IN VITRO*

G. Takagaki

Department of Physiology, Keio University School of Medicine, Tokyo.

It is known that the addition of glutamic acid to the incubation medium increases aerobic glycolysis in cerebral cortex slices. Ouabain (10^{-5} M) inhibited the effect of glutamic acid. When potassium ions were omitted from the medium, or when they were added to the medium at a higher concentration than 40 mM, glutamic acid was not effective in the promotion of aerobic glycolysis. These findings show that the effect of glutamic acid is evoked by the potassium ions accumulated at a high concentration in the slices. Analysis of intermediates of glycolysis in slices incubated in the presence as well as in the absence of a high concentration of potassium ions, indicated that the pyruvate kinase step is one of the rate-limiting points in the reactions of aerobic glycolysis, and that this step is activated by the addition of a high concentration of potassium. The pyruvate kinase from brain tissue was very strongly inhibited by a low concentration of calcium ions and activated (or deinhibited) by potassium ions. The activity of the pyruvate kinase was increased by potassium ions proportionally to their concentration up to 150 mM. These properties of the pyruvate kinase were found to be characteristic of the enzyme from brain tissue, and this may explain the mechanism of the effect of potassium ions in the slices. The addition of aspartic acid to the medium promotes the accumulation of potassium ions in the slices, as the addition of glutamic acid does. However, the effect of glutamic acid on aerobic glycolysis was inhibited by aspartic acid; its mechanism of action may be explained by the antagonistic effect of calcium and potassium upon the pyruvate kinase. These observations are to be discussed in relation to the unique regulation mechanism of glycolysis in brain.

RNase AND RNase INHIBITOR, EXAMINED IN RELATION TO PROTEIN AND RNA METABOLISM IN THE CENTRAL NERVOUS TISSUE

Y. Takahashi and Y. Suzuki

Brain Research Institute and Department of Neurology and Psychiatry, Niigata University School of Medicine, Niigata, Japan.

Free acid RNase, free alkaline RNase, total acid RNase and total alkaline RNase were found in brain tissue homogenate and subcellular particles. High RNase activity was found in rabbit hippocampus, cerebral cortex and cerebellum, while the cerebral white matter and spinal cord showed low values. The supernatant fraction of brain tissue contained a RNase-inhibitor. Regional distribution of the RNase inhibitor in the rabbit brain was similar in pattern to that of RNase. The activities of RNase and RNase inhibitor in the developing rat cerebral cortex were also estimated.

In further experiments, RNase inhibitor was purified from pig brain. The homogenate of cerebral cortex was centrifuged at 105000 g for 90 min to obtain the supernatant fraction, which contained RNase inhibitor. Further purification procedures of RNase inhibitor involved: (1) DEAE cellulose column chromatography, (2) hydroxylapatite column chromatography and (3) sephadex G-100 gel filtration. The RNase inhibitor activity of the most active fraction eluted from sephadex G-100 gel filtration could be purified about 6000-7000 fold. Analysis of this active fraction by disk electrophoresis showed that a major band represented about 80-85% of total protein. This active fraction inhibited pancreatic RNase, brain microsomal and mitochondrial alkaline RNase, but did not inhibit acid RNase and RNase T_1. The purified RNase inhibitor was inactivated by a low concentration of PCMB, and by preincubation with certain proteolytic enzymes.

Previous report (TAKAHASHI, MASE & ABE, 1966) showed that rat brain ribosome was able to incorporate [^{14}C]leucine into protein. An attempt was made to prepare polysomes as the functionally active form of ribosomes, but, it was found difficult to prepare polysomes from rat brain tissue without addition of the RNase inhibitor fraction to the preparation medium. The amino acid incorporating activity of brain polysomes was higher than that of brain ribosomes (TAKAHASHI, MASE & SUGANO, 1966). The possibility will be discussed that RNase and RNase inhibitor play a role in protein biosynthesis in the brain.

TAKAHASHI, Y., MASE, K. & ABE, S. (1966). J. Biochem., 60, 363.
TAKAHASHI, Y., MASE, K. & SUGANO, H. (1966). Biochim. biophys. Acta, 119, 627.

QUALITATIVE AND QUANTITATIVE PATTERNS OF GANGLIOSIDES IN THE BRAINS OF DIFFERENT ANIMALS

G. Tettamanti and V. Zambotti

Istituto di Chimica Biologica dell'Universitá di Milano, Milano

The method of preparing gangliosides set up in our laboratory was used successfully for the isolation, identification and semi-quantitative determination of the different gangliosides (up to 12) present in the brain of shark, frog, chicken, ox, pig, horse, rat, rabbit, Macacus rhesus, Cercopitecus ana man.

The qualitative pattern of gangliosides was found to be quite similar in the animals tested from chicken to man (gangliosides $G_{GNT}I$, $G_{GNT}II$, $G_{GNT}III$ and $G_{GNT}IV$ make up approximately 80-90% of the total ganglioside content; gangliosides $G_{GNT}V$, $G_{Tr}I$ and G_{Lact} are present in small amounts). A different pattern was found in shark brain (the gangliosides having a short oligosaccharide chain are predominant: about 70% of the total) and frog brain (the tetrasialoganglioside $G_{GNT}V$ makes up more than 50% of the total).

The quantitative pattern was found to be similar only in the animals of similar species (for instance ox, pig, horse or Macacus rhesus, Cercopitecus and man).

Typical gangliosides were isolated from frog, chicken and pig brain (a tetrahexosepentasialoganglioside from frog and chicken brain; a tetrahexosedisialoganglioside from pig brain). During the present study several gangliosides of unknown structure were isolated. The chemical structure of these new gangliosides is under investigation.

MYELINOGENESIS AND BRAIN FUNCTION IN PRENATALLY X-IRRADIATED RATS

Paola S. Timiras and Antonia Vernadakis

Department of Physiology,
University of California, Berkeley

To determine whether the earlier appearance of seizure activity in prenatally X-irradiated rats is accompanied by accelerated myelination of developing nerve fibres, brain myelinogenesis was investigated after prenatal radiation. Pregnant rats were exposed to a single dose of 100R whole-body X-radiation at 14 days of gestation. Lipid and glycolipid concentrations were measured in the cerebral cortex and diencephalon plus midbrain of 36 control and 36 irradiated offspring at 9, 12 and 16 days after birth. Lipids were extracted by the FOLCH-PI method, determined gravimetrically, and expressed as mg/100 mg wet tissue. Glycolipids were obtained through chromatography, determined by galactose content, and expressed as mg/100 mg lipid. Body weights of irradiated offspring were the same, whereas brain weights were lower than those of controls. Lipid in the cerebral cortex and diencephalon plus midbrain increased at the same rate in all animals. From 9 to 16 days of age, glycolipid concentration in the diencephalon plus midbrain increased from 1.65% to 6.69% in controls, and from 2.56% to 7.67% in irradiated animals; in the cerebral cortex it increased from 1.01% to 1.77% in controls, and from 1.17% to 2.09% in irradiated animals. The rate of increase in these subcortical areas is considerably more rapid than in the cortex. Glycolipid concentration was higher in the diencephalon plus midbrain, but not in the cerebral cortex of irradiated offspring. Morphological and chemical studies have shown that the number of brain glial cells increases after prenatal X-radiation. Glial cells are considered to be involved in myelinogenesis in the CNS. Glycolipid changes, therefore, may be correlated with the increase in number of glial cells. The question of whether the increase in glycolipid concentration represents myelination of axons of developing neurons, or accumulation of non-functional, intracellular lipid material remains to be answered.

(Supported by AEC)

THE INFLUENCE OF NUCLEASES, AND INHIBITORS OF NUCLEIC ACID AND PROTEIN SYNTHESIS, ON BEHAVIOUR IN THE MOUSE

W. Tongur and O. Krylov

Laboratory of Nucleic Acids, Institute of Biological and Medical Chemistry, Academy of Medical Sciences, Moscow, and Institute of Brain Research, Academy of Medical Sciences, Pereulok, Obuhar 5 (USSR)

STUDIES ON RNA SYNTHESIZED *IN VITRO* BY SENSORY GANGLIA OF CHICK EMBRYO

G. Toschi, E. Dore and O. Sapora

Institute of General Physiology, University of Rome, and Istituto Superiore di Sanità, Rome.

RNA labelled with radioactive uridine or with ^{32}P was examined by chromatography on columns of methylated albumin and by centrifugation in sucrose gradients. Peaks were analysed for nucleotide composition. Labelled RNA was examined in relation to different times of labelling and to the action of a specific growth factor (NGF).

MULTIPLE SCLEROSIS BRAIN GAMMA GLOBULIN AND THE SINK ACTION OF THE CEREBROSPINAL FLUID

W.W. Tourtellotte

Department of Neurology,
University of Michigan Medical School, Ann Arbor, Mich.

This report will present data to support the hypothesis that some of the immunoglobulin-G (IgG) in cerebrospinal fluid (CSF) can originate from the brain in patients with multiple sclerosis (MS) who have elevated CSF IgG values.

Unfixed serial coronal cerebral sections (3 mm thick) were made about 24 hours after death, sealed in plastic bags and stored at -90°C. Coronal sections which included at least one ventricle from 12 patients who died with multiple sclerosis and from six patients who died with no detectable (gross or microscopic) structural central nervous system disease (controls) were thawed and grossly dissected especially around the ventricles along anatomical-pathological lines into either normal appearing white matter and/or plaques of demyelination. The dissected pieces were analyzed histologically in order to determine the activity of the demyelinating process and immunochemically for albumin and IgG. The latter determinations were corrected for residual blood IgG and albumin found in the brain tissue. The IgG and total protein of the CSF which was obtained approximately six months prior to death was also determined for each MS patient.

A positive correlation was found between the concentration of IgG in periventricular MS brain tissue and the concentration of IgG in the CSF. Since the concentration of the tissue IgG was higher than the premortal CSF values in all the MS cases studied, it is reasonable to assume that the increased tissue IgG content underlies the increased concentration in the CSF. However, the relationship is probably not a simple one, since the data showed that the CSF values did not exceed 36 per cent of the total protein as IgG, although the brain concentration may be further increased approximately six times (411 to 2400 mg IgG/kg of brain). This could mean that the homeostasis of CSF IgG at least in part is the result of a dynamic equilibrium such that IgG enters the CSF faster than it leaves up to a certain point (36 per cent IgG).

Based on these data it would appear that the CSF IgG value from a patient with MS can be used to predict with a high degree of probability ($p \leq 0.003$) that the brain IgG concentration particularly around the ventricles will be above a cut off point of 411 mg IgG/kg of tissue (average of control white matter plus 3 standard deviations).

Furthermore, the concentration of the IgG in the MS brain and the degree of immunofluorescence staining of the

MS brain tissue appeared to be related to the histological activity of the demyelinating process which was characterized by microglial and astroglial proliferation, as well as perivascular mononuclear infiltration.

To further validate this hypothesis we have attempted to destroy the multifocal areas of synthesis of IgG in the brain by x-irradiation (600 R) of the head of a MS patient whose CSF IgG was significantly elevated ($p \leq 0.003$). The results of five weekly CSF IgG values obtained pre-x-irradiation will be compared to weekly post-x-irradiation values. The significantly lowered CSF IgG values post-x-irradiation offer further support that the source of the elevated CSF IgG in MS patients is probably due to multifocal areas of synthesis of IgG in the brain itself.

This work was supported in part by the USPHS, NIH NB-05388-03 and the National Multiple Sclerosis Society 272-6).

OUABAIN AND ELECTROLYTE METABOLISM OF CEREBRAL CORTEX IN VITRO

Donald B. Tower

Laboratory of Neurochemistry,
National Institute of Neurological Diseases and Blindness,
Bethesda, Maryland.

At concentrations of 10^{-5} M or higher, ouabain inhibits the extrusion of Na and accumulation of K by slices of cerebral cortex incubated under otherwise optimal aerobic conditions in vitro (JOANNY & CORRIOL, 1964; SWANSON & MCILWAIN, 1965; BOURKE & TOWER, 1966). The effect on metabolism of monovalent cations is much greater than concomitant effects on slice swelling or distribution of sucrose tracers (BOURKE & TOWER, 1966). In media containing Ca, respiratory metabolism of ouabain-inhibited slices is stimulated, whereas inhibition is observed in Ca-free media (SCHWARTZ, 1962; SWANSON & MCILWAIN, 1965; WHITTAM *et al*., 1965; BOURKE & TOWER, 1966).

We have now investigated the possibility that in incubated slices of cerebral cortex, ouabain also affects Ca metabolism, as in the case of cardiac muscle (GOVIER & HOLLAND, 1964, 1965; KEETON & BRIGGS, 1965). Slices of cat cerebral cortex were incubated in bicarbonate-saline-glucose media containing either 5 or 27 mK-K and ± Ca in the presence or absence of 10^{-5} M-ouabain. Slices and media were ashed at 550° and analysed for Ca and Mg by atomic absorption flame spectrophotometry (Perkin-Elmer 303). The Mg content of such incubated slices was unaffected by the presence of ouabain, regardless of whether or not Ca was included in the media. In Ca-containing media, inhibitory concentrations of ouabain were associated with a pronounced increase (about 50%) of the Ca content in the slices. In Ca-free media, slice levels of Ca were minimal (about 20% of controls), as previously reported by LOLLEY (1963). Data are now being obtained from incubated slices of cat corpus callosum and of neonatal kitten cerebral cortex for comparison with data from adult preparations.

Since comparable concentrations of ouabain when infused intracranially *in vivo* produce leakage of K into CSF and severe convulsions (CSERR, 1965; KATZMAN, 1965; BIGNAMI & PALLADINI, 1966) as well as characteristic neuropathology (BIRKS, 1962; BIGNAMI & PALLADINI, 1966), studies with ouabain may provide useful model systems for elucidating factors contributing to stability and lability of conducting membranes. The key role of Ca both in the membrane (TOBIAS, 1964) and in mitochondrial metabolism (CHANCE, 1966; SLATER, 1966) deserves special emphasis.

BIOCHEMICAL STUDIES ON ORGAN-SPECIFIC PROTEIN FROM BRAIN MICROSOMES

Y. Tsukada, K. Uyemura and T. Noguchi

Department of Physiology, Keio University School of Medicine, Tokyo.

Biochemical studies of membrane structures from nervous tissues were attempted.

The microsomal fraction was prepared from rat brain homogenates by differential centrifugation. The fraction was treated by EDTA or DOC to remove ribosomes from the membrane fragments and antisera were obtained from rabbits injected with the treated microsomal fraction. When the antisera were allowed to react with antigen on agar diffusion plates, two bands of antigen-antibody precipitate were clearly observed. However, the microsomal fraction from liver, kidney, or from serum proteins of the rat, did not react as an antigen against anti-brain microsome globulin. Also antisera did not give immunological reactions with brain microsomal extracts from beef, rabbit, guinea pig or chick. This means that brain microsomes contain organ-specific proteins which are also species-specific.

It was found that the antibody of rat brain microsomes inhibited the Na-K-ATPase localized in the brain microsomal fraction.

The purification of the antigenic microsomal protein was performed. When microsomal fractions were extracted with 10 vol deionized water, and sonic oscillation applied (20 kc 150 w for 5 min in the cold), the combined supernatant obtained by 105,000 g centrifugation for 60 min and repeating the procedure 3 times contained over 90% of the protein, RNA, DNA and phospholipids. The antigenic protein of the microsomal fraction extracted by this method could be further purified by DEAE cellulose chromatography, acrylamide gel disc electrophoresis and $(NH_4)_2SO_4$ fractionation. Each fraction separated was checked chemically.

The antigenic protein appeared in the precipitate obtained with 20% and 80% sat. $(NH_4)_2SO_4$. The latter moved to the negative side on immunoelectrophoretic examination, and the former kept the original position. From the chromatographic examination, it appeared that the organ-specific protein of the microsomal fraction is basic in nature. The physicochemical properties of these proteins are being examined.

H

GLUCOSE BALANCE AND SYNTHESIS OF FREE AMINO ACIDS IN THE CNS OF NORMAL AND GAMMA(Co^{60})-IRRADIATED RATS

Ramon Valderas, Norma Francotte and Enrique Egana

Institute of Experimental Medicine, School of Medicine, University of Chile, Santiago.

The synthesis of free amino acids in the CNS is closely coupled - among other mechanisms - with the supply of glucose from the cerebral blood stream and also with the actual glucose consumption taking place in the nervous tissue. In the present work [1-^{14}C]glucose was used. Rats were exposed to a low 50 r whole body gamma irradiation with Co^{60}. The effect on the synthesis of free amino acids and the glucose content of the brain and also the arteriovenous glucose difference were studied. Various times after the gamma exposure and the intraperitoneal injection of the [1-^{14}C]glucose (15, 30, 60 and 120 minutes) were selected. In all experiments the specific radioactivity and content both of the free amino acids (nine in number) and glucose were determined. In general it may be said that there appeared to be a temporarily increased synthesis of free amino acids, although some amino acids (glutamic, aspartic, glutamine) showed a decrease in the actual content, measured chemically. The specific activity and content of glucose in the arterial and venous blood did not show significant changes that could explain these results on free amino acids. The CNS glucose content decreased significantly between 30 to 60 minutes after the exposure.

Stimulations by various different methods, such as pricking, produced the same effect on glucose content. It is of interest to note that GABA behaved in a different way from the other amino acids.

MOLECULAR PROPERTIES OF THE NERVE GROWTH FACTOR

Silvio Varon, J. Nomura and E.M. Shooter

Department of Genetics,
Lt. Joseph P. Kennedy, Jr. Laboratories for Molecular Medicine,
Stanford University School of Medicine,
Palo Alto, California.

The Nerve Growth Factor (LEVI-MONTALCINI, 1966) is a protein which has been isolated from snake venom and the submaxillary gland of the adult male mouse with reported molecular weights of 20,000 and 44,000 respectively (COHEN, 1959; 1960). A new form of the NGF protein has been obtained in considerably increased yield from the mouse submaxillary gland. It has the same specific activity as the original preparation but an $S_{20,w}$ of 7.1 corresponding to an approximate molecular weight of 140,000. This NGF species is stable between pH 5 and 8, but dissociates reversibly outside these pH limits. At pH 3.8 and 10.3 the dissociation is complete and all the subunits, as well as the NGF activity, sediment together in a single boundary with an $S_{20,w}$ of 2.5. Only about 25-35% of the original activity is recovered at extremes of pH. Virtually complete recovery of both the high molecular weight protein and the original NGF activity is achieved when acid or alkali treated material is returned to neutral pH.

Chromatography on CM-cellulose at acid pH resolves three groups of NGF subunits, one group being sufficiently acidic to be eluted at this pH. The acidic group and a second group eluted at neutral pH are inactive. NGF activity is found exclusively with the third most basic group of subunits; again only 25-35% of the original activity is recovered in this fraction. The isolated subunits do not aggregate at neutral pH. Recombination of the active fraction with either inactive fraction results in a shift of activity to higher molecular weights but does not prevent the decrease of activity noted for the active subunit alone. Recombination with both inactive fractions increases the total NGF activity and reproduces a species with the sedimentation, electrophoretic and activity characteristics of the high molecular weight NGF. The properties of the various subunits and of their recombination products will be discussed.

COHEN, S. (1959). J. biol. Chem., 234, 1129. COHEN, S. (1960). Proc. nat. Acad. Sci. (Wash.), 46, 302.
LEVI-MONTALCINI, R. (1966). Harvey Lectures, 60, 217.

GLYCEROLPHOSPHATE DEHYDROGENASE DIFFERENTIATION IN THE DEVELOPING RAT BRAIN: EFFECT OF X-IRRADIATION

J. de Vellis

Laboratory of Nuclear Medicine and Radiation Biology,
U. C. L. A. School of Medicine,
University of California, Los Angeles

Isoenzymes of glycerolphosphate dehydrogenase (E.C.1.1.1.8) were isolated by cellulose acetate electrophoresis. Form III was predominant in the mature brain. The development of III was observed to occur between 18 and 45 days post partum in several regions of the rat brain. On the other hand, Form II was the major band up to 20 days of age.

The rise in activity of III occurs at a time where cell division has ceased. It is therefore possible to study the effect of ionizing radiation on enzyme synthesis in the absence of cell division, which is one of the most radiosensitive processes. A dose of 750r administered to the heads of 20-day old rats inhibited as much as 50% of the development of III. The dose-response curve of this inhibition was sigmoid in form (from 90r to 1500r). The inhibition of the rise in activity of III is a delayed phenomenon, probably owing to the presence of long-lived mRNA as evidenced by studies with actinomycin D, but no recovery of the inhibition is finally observed. A more immediate phenomenon with no recovery was a 10% degradation of DNA taking place during the first 24 hours following irradiation. By comparison with III, the activity of enzymes known to occur in glial and neuronal cells remained unchanged. Fully developed III was not affected by irradiation. These results suggest an x-ray induced inhibition of the genetic transcription for GPDH III and that there is no DNA repair mechanism in the brain comparable to the one observed in E. coli following irradiation. Preliminary experiments indicate that Form II, and maybe early development of III, might be more directly involved in the synthesis of myelin phospholipid than late Form III.

This work was supported by the USAEC.

SEPARATION OF THE LOW MOLECULAR WEIGHT PEPTIDE FRACTION FROM CALF BRAIN STEM AND IDENTIFICATION OF SOME ACIDIC COMPOUNDS

D.H.G. Versteeg and A. Witter

Department of Pharmacology, University of Utrecht, Utrecht, The Netherlands

Gel filtration on Sephadex G25 of a 0.1 N-hydrochloric acid peptide extract from an acetone-dried powder of calf brain stem, resulted in the separation into 5 fractions. This paper will present data on the further fractionation of the low molecular weight fraction E, consisting of amino acids, amino acid derivatives and small peptides. Separation of fraction E on Dowex 50 x 4, applying linear gradient elution, yielded about 16 different fractions. Attention was first of all focussed on those fractions that appeared in the eluate before aspartic acid. The most acidic fractions eluted from the Dowex 50 column, were further separated on Dowex 1 X-4. The compounds thus obtained were further investigated by paper chromatography and paper electrophoresis.

Besides classical analytical methods, mass-spectrometry was employed in elucidating the structure of these acidic peptides and amino acid derivatives. The structures of compounds identified so far will be given.

MODIFICATION OF THE POLYSOMAL PATTERN IN RABBIT CEREBRAL CORTEX BY ELECTROSHOCK TREATMENT

C. Vesco and A. Giuditta

International Laboratory of Genetics and Biophysics, Naples

There is considerable evidence that RNA and protein metabolism in nervous tissues may be significantly modified by changes in functional state. Intense excitation has generally been found to cause an increase of RNA and protein catabolism. In the present work the influence of hyper-activity was investigated on the polysomal content of rabbit cerebral cortex. The occurrence of a dynamic equilibrium between free ribosomes and ribosomal aggregates, which is sensitive to functional changes, has been observed in several cell types (bacteria, antibody-producing cells, regenerating liver). Massive stimulation, in view of the relative insensitivity of biochemical analysis, was provided by a series of epileptic seizures induced by electric shocks. Under these conditions the equilibrium between ribosomes and polysomes was shifted with a considerable increase (approximately 100%) in the content of free monomers at the expense of ribosomal aggregates. The shape of the polysomal profile remained essentially unmodified, as expected from a proportional reduction in the amounts of all the classes of aggregates. Control experiments suggested that these modifications were not due to an activation of nucleases, but probably represented the net result of a new steady state attained between the rates of formation and disaggregation of polysomes.

FINE STRUCTURE AND HISTOCHEMISTRY OF PERIPHERAL BLOOD LEUCOCYTES AND PLATELETS IN NIEMANN-PICK DISEASE

Bruno W. Volk, Sydney S. Lazarus and Larry Schneck

Isaac Albert Research Institute, Brooklyn, New York

Leucocytes and platelets in the buffy coat of peripheral blood from four children with Niemann-Pick disease, five Tay-Sachs patients, and three non-lipidosis patients were studied by optical microscopy and the electron microscope after double fixation in glutaraldehyde and osmium tetroxide. By optical microscopy, about 75% of the lymphocytes contained 1 to 9 vacuoles in the cytoplasm, while they were present in only 4% of Tay-Sachs patients. Ultramicroscopically, the majority of lymphocytes of the Niemann-Pickpatients contained lipid cytosomes at the periphery of the Golgi zone and occasionally at other sites. These inclusions were composed of concentric or parallel arrays of membranes with a single limiting membrane. Frequently they showed also finely granular material and small vesicles. Monocytes also contained cytosomes at the periphery of the Golgi zone but smaller than those in lymphocytes. Some lymphocytes and monocytes also contained an accumulation of clear vacuoles at the cell surface apparently for discharge to the exterior. Neutrophils did not contain any lipid cytosomes and the granules were normal. Basophils contained lipid bodies in addition to their specific granules which were unaltered. Some eosinophils showed small whorls of membranes inside specific granules. Platelets also occasionally exhibited memgranous bodies. RBC's were unchanged. An apparent sequence of formation of cytosomes from Golgi derived finely granular bodies could be traced. With the electron microscope, acid phosphatase activity was demonstrated at the inner surface of the limiting membranes, and at the periphery of contained vesicles of cytosomes as well as in Golgi elements. This suggests that lysosomal enzymes present in Golgi derived granular bodies transforms them into lipid bodies. In tissue culture the lymphocytes of Niemann-Pick patients also contained similar lipid cytosomes. These findings taken in conjunction suggest that the cytosomes within WBC's are of endogenous origin. Neither Tay-Sachs nor non-lipidosis patients showed lipid bodies in their blood constituents.

BIOSYNTHESIS OF RNA IN 'NEURONAL' AND 'NEUROGLIAL' FRACTIONS

P. Volpe and A. Giuditta

International Laboratory of Genetics and Biophysics, Naples

The occurrence of a complex interrelationship in the RNA metabolism of glia and neurons has been repeatedly suggested on the basis of experiments carried out by microanalytical techniques. The availability of methods for the bulk preparation of 'neuronal' and 'neuroglial' cell fractions has prompted an investigation of some aspects of this problem by more conventional biochemical procedures. Cell fractions were prepared from rabbit cerebral cortex by the method of ROSE and in separate experiments by the method of SATAKE and ABE. Extraction of RNA from these fractions was carried out by a hot phenol-SDS procedure at different times after the subarachnoidal injection of [^{14}C]orotic acid. Sedimentation analysis was performed by centrifugation in sucrose density gradients. At early times after labelling (one hour) the specific activity of total 'neuronal' RNA was significantly lower than that of the 'neuroglial' fraction, but after three and six hours it became considerably greater. At still later times (14 hours) the ratio of specific activities in the two fractions approached unity. These differences were most marked for the heavy species of RNA (18 S, 28 S, > 28 S) and were less evident for the light types of RNA (4 S). Substantial agreement was found in the kinetic of labelling of 'neuronal' RNA prepared by the methods of ROSE and SATAKE and ABE.

BRAIN TRANSGLUTAMINASE DURING DEVELOPMENT OF EAE

I.J. Wajda and J.M. Lee

New York State Psychiatric Institute and
Rockefeller University, New York

The antigen of autoimmune disease has been thought to be a tissue component, most likely a protein sufficiently altered to become a heterologous substance. Transglutaminase, an enzyme which catalyzes the exchange of amide groups of protein bound glutamine with primary amines, or with the ε-amino group of protein bound lysine, could play a major enzymic role in changing body proteins into antigenic determinants. Transglutaminase is a cellular enzyme, assayed in the soluble fraction of the centrifuged tissue homogenates. It can be found in practically every organ, except plasma or serum the liver being the richest source. It is known that in autoimmune disorders, also when they are produced experimentally, the genetic predisposition plays a very important role. Two genetically pure strains of mice were used in our experiments: BSVS strain, which easily develops experimentally produced allergic encephalomyelitis (EAE) and BRVR genotype which is resistant to the disease.

During the development of EAE the levels of transglutaminase were followed first in the livers of experimental animals. The enzyme was assayed using a colorometric method based on hydroxamic acid formation. We were able to show that the changes in liver transglutaminase preceded the development of the disease, and that the sensitive strain of mice responded with significantly higher levels of the enzyme when compared to the resistant strain. It was of course of primary interest to establish if the enzyme changes could be detected in the brain. A sensitive method of transglutaminase assay was developed, using a synthetically prepared substrate, which in the presence of transglutaminase accepts radioactive amine. Using the same two genetically pure strains of mice, we repeated our previous experiments, and followed transglutaminase levels in the brains of injected mice during the development of EAE. In each experiment, two groups of 100 mice from both strains were injected, six animals from each group were sacrificed on consecutive days; brain and liver transglutaminase were assayed. The results of this investigation will be presented and discussed.

INFLUENCE OF SOME IONS AND SMALL MOLECULES ON THE ACTIVITY AND PHYSICAL PROPERTIES OF A PURIFIED ANIONIC ASPARTIC-2 OXOGLUTARIC AMINOTRANSFERASE: A MODEL SYSTEM

A. Waksman

Centre de Neurochimie du C.N.R.S., Faculté de Médecine, Strasbourg.

Glutamic acid (and its metabolic products) plays a central role as a modulator in the central nervous system (CNS). Aspartic-2 oxoglutaric aminotransferase (GOT) has the double advantage of being part of the glutamic acid pathway, and of existing in multiple isoenzymic forms, which are separable by electrophoresis into groups (cationic and anionic) each of which belongs to a defined biological compartment (respectively mitochondrial and extramitochondrial). This confers on the GOT system the value of a model system for the study of the differential regulation of the GOT activities.

This study was performed using as a model in the first place a highly purified anionic GOT preparation from pig heart. Our observation showed that such a preparation was weakly active in NaCl, KCl, NaBr and NaF, whereas Na_2HPO_4, K_2HPO_4, Tris and its aminoalcohol derivatives, ethanolamine, glucosamine and galactosamine had a very marked activating effect on the enzyme. All the alcohols, polyols and amines tested, and N-acetylglucosamine, N-acetylgalactosamine, phospho-ethanolamine, serine, phosphoserine, ATP, taurine, acetyl-aspartate and gangliosides, as well as some active transmitters at the CNS level, were very weakly effective or without influence at all on the activity of anionic GOT. No change in the molecular weight could be detected during enzymic activation, but the existence of binding sites for cations and anions is suggested by the electrophoretic and kinetic data.

From our results, a relationship between the regulation of the anionic-extramitochondrial GOT activity and the membrane-bound ATPase activity might be suggested as a working hypothesis.

The study of the differential behaviour of the anionic and cationic pig brain GOT's is presently under investigation.

PROPERTIES OF A "BRAIN-SPECIFIC" GLYCOPROTEIN IN AQUEOUS EXTRACTS OF HUMAN CENTRAL AND PERIPHERAL NERVE TISSUE

Krystyna Warecka and H. Bauer

Neurochemisches Labor, Nervenklinik der Universität, Göttingen, Germany.

The water-soluble "brain-specific" protein migrating in the $alpha_2$ area of agar and agarose media is a heterogeneous (double line, divergence) component. It is strongly PAS-positive and experiments with neuraminidase indicate that it contains neuraminic acid. It is a constant component of white and grey matter of brain, spinal cord and peripheral nerve tissue (ulnar nerve, sciatic nerve, cauda equina, plexus sacralis). The brain-specific protein is present in normal and pathological brain tissue (3 multiple sclerosis, 1 leuco- or panencephalitis, 1 meningoencephalitis, 1 encephalopathy of undetermined origin). The fraction could not be found in fetal brains or in the brains of the newborn; it was present in the brains of two-year old infants, however (over 30 brains). By ultracentrifugation, localization in microsomes could be demonstrated immunologically. A cross reaction with rat brain takes place. As an antigen the water-soluble "brain-specific" glycoprotein is probably not encephalitogenic.

THE INTRACELLULAR DISTRIBUTION OF BRAIN CATECHOLAMINES

H. Weil-Malherbe and Y. Gutman

National Institute of Mental Health,
St. Elizabeths Hospital, Washington, D.C.

Tissue catecholamines, as is well-known, are distributed between a soluble and particle-bound fraction. Although the existence of a soluble fraction may be postulated on theoretical grounds, the true *in vivo* concentration of catecholamines may be much smaller than that usually found in tissue homogenates.

Experiments were performed to assess the importance of artefacts in the distribution of norepinephrine and dopamine in brain homogenates. Reducing the time interval between death and the preparation of the homogenate significantly increased the particle-bound portion of norepinephrine and dopamine. Rapid cooling further increased the particle-bound fraction of norepinephrine but had no effect on the distribution of dopamine. The dilution of the homogenate, the size of the animal or anaesthesia before decapitation did not affect the distribution of either catecholamine. Relatively more norepinephrine than dopamine was released from suspended particles by freezing and thawing or by incubation at 30°. When whole homogenates were incubated, the rate of release was increased and was the same for norepinephrine and dopamine.

The intracellular distribution of norepinephrine and dopamine is different in different areas of rabbit brain. Of the norepinephrine in the hypothalamus 75% was particle-bound whereas of the dopamine in the basal ganglia only 40% was particle-bound.

No evidence was found for the existence of a stable complex between norepinephrine and protein in the soluble fraction of rabbit hypothalamus.

It is concluded that much of the soluble fraction, especially of brain norepinephrine, is due to artefact and that norepinephrine and dopamine differ in the extent and firmness of their combination with binding sites in the brain.

ELECTROLYTES OF THE BRAIN IN WATER INTOXICATION

M. Wender and W. Strzyzewski

Department of Neurology, Medical Academy, Poznań, Poland

Investigations were carried out to define the changes in the water and electrolytes of the brain during water intoxication, which was induced in the white rat by oral administration of 50 ml of water simultaneously with intramuscular injections of pitressin.

We found that the increase in the water content of the cerebral cortex during these experiments was due mainly to accumulation of water in the 'chloride space', which represents the extraneuronal compartment. The changes of water content in the cerebral cortex were parallel with the decrease in absolute values of sodium and chloride expressed in mEq/g of fresh tissue. In comparison with the serum values, the concentrations of sodium and chloride in the cortex were clearly increased. The data on potassium were not statistically significant.

The changes in water and electrolytes in the brain found in the syndrome of water intoxication are similar to those observed in experimental edema. However we found no significant differences between the groups of animals, which did and did not have seizures during the experimental procedure.

In both experimental groups the water content of the cerebral cortex increased in parallel with the increase in hydration of the red blood cells, which may reflect the general trend of the alteration in water metabolism in other tissues.

INCREASED RNA SYNTHESIS IN THE BRAINS OF MICE UNDERGOING CONDITIONED AVOIDANCE TRAINING

John Eric Wilson, John W. Zemp and Edward Glassman

Department of Biochemistry, University of North Carolina School of Medicine, Chapel Hill, North Carolina

The influence of a conditioned avoidance training experience on the incorporation of radioactive uridine into the RNA of mouse brain was investigated. A double isotope labelling method was used in which one mouse of a pair was injected with [^{14}C]uridine into the frontal lobe to a depth of 3 mm while the other mouse was injected with [^{3}H]uridine. Thirty minutes later, one of the pair was randomly chosen to receive 15 minutes of training to avoid an electrical shock. The remaining animal of the pair served as a yoked control. The animals were sacrificed immediately after training and the brains were quickly removed in the cold. Nuclei and ribosomes/polysomes were separated from a homogenate of the combined brains, and RNA was isolated from each fraction by a phenol extraction procedure. The RNA from nuclei and from the ribosome/polysome fractions isolated from the brains of trained mice showed greater isotope incorporation than RNA from yoked controls. Sucrose density gradient centrifugation of the nuclear or the ribosomal/polysomal RNA yielded isotope sedimentation patterns similar to the patterns of equivalent fractions of rapidly labelled RNA from liver reported by other workers. The increase in radioactivity in the RNA from the nuclei of the trained animal was distributed throughout the gradient and not in any one peak, indicating that the increase is in a population of RNA molecules of diverse sizes. The increase in radioactivity of RNA from the ribosomes of the trained animal was also distributed throughout the gradient but tended to be greatest at about 18 s. In contrast to the RNA of brain, there was no increase of radioactivity in the RNA from the livers or the kidneys of the trained animals suggesting that we are dealing with processes specific to the brain. Since studies of free nucleotides in the brain indicate that permeability changes cannot account for these data, we conclude that the conditioned avoidance training is accompanied by an increased synthesis of rapidly labelled RNA of the brain. It is difficult, however, to ascribe this effect to a specific aspect of the training experience. Learning is only one difference between the trained and control mice, and such factors as stress, organized locomotion, attention, etc., must be considered as possible causative agents.

EVALUATION OF EXTRACTION PROCEDURES AND PRELIMINARY SEPARATIONS OF PEPTIDES FROM CALF BRAIN STEM

A. Witter and D.H.G. Versteeg

Department of Pharmacology, University of Utrecht, Utrecht, The Netherlands

Different extraction procedures for obtaining peptides in high yield in a mild and reproducible way from an acetone-dried powder of calf brain stem were compared. The evaluation of the different methods was based on the colour yields of the extracts with ninhydrin, together with the increase of these colour yields after alkaline hydrolysis of the extracts. The results, obtained with some twenty different solvent systems and including pre-extractions of the acetone-dried powder with lipid solvents, will be discussed in the light of these two parameters. The interpretation of the results led to the adaptation of an extraction procedure that involved chloroform in the pre-extraction of lipids, followed by an extraction of peptides with 0.1 N-hydrochloric acid.

Gel filtration of the final extract on Sephadex G25 resulted in a highly reproducible separation into 5 fractions. The yields and nature of these fractions will be discussed.

The acetone extract, obtained in the preparation of the acetone-dried powder, was investigated for peptides, that might possibly be present in this extract. Separations on Sephadex G25 and Sephadex G10 were used in order to gain information on the nature of the ninhydrin-positive substances present in the acetone extract. An interpretation of the results will be presented.

PROTEINS OF CENTRUM OVALE, OPTIC NERVE AND PERIPHERAL NERVE MYELIN

Frederick Wolfgram

Department of Medicine, Division of Neurology, U.C.L.A. School of Medicine, Los Angeles, California

Purified myelin was isolated from human, monkey, and bovine centrum ovale white matter and from bovine optic nerve. Bovine intradural spinal roots were used for isolating peripheral nerve myelin. The purity of these preparations was examined by electron microscopy. Centrum ovale myelin was almost completely soluble in chloroform-methanol but peripheral nerve myelin proteins were insoluble in this solvent. The proteins of both types of myelin were freed of lipid by denaturation from chloroform-methanol-water and then were digested with trypsin. Data will be presented on the amounts and amino acid composition of the trypsin-resistant protein in each type of myelin. The amino acid composition of the three myelins was determined after both 22 and 70-hour hydrolysis. In general, the amino acid patterns of the centrum ovale myelin of the three species were remarkably similar. However, differences were apparent between bovine centrum ovale and peripheral nerve myelin.

The basic proteins that are soluble in dilute acid were extracted from the bovine centrum ovale, optic nerve and peripheral nerve myelin after removing the bulk of the lipids by extracting first with chloroform and then with methanol. Acrylamide gel electrophoresis demonstrated one major basic protein band in CNS myelin and several very minor bands. The major band comprised about 90% of the total CNS myelin basic proteins and it is the antigen for allergic encephalomyelitis. The basic proteins of peripheral nerve consisted of four bands of almost equal amount and these had mobilities different from that of the major band of CNS myelin.

ACTIVITY OF MEMBRANE-BOUND ENZYMES AS A FUNCTION OF STRUCTURAL CHANGES IN MEMBRANES

M. Wolman

Department of Pathology, Histology and Cytology,
Tel-Aviv University Medical School, Government Hospital,
Tel-Hashomer, Israel.

In closely apposed membranes, as in the intercellular spaces of the CNS and to a higher degree between the lamellae of the myelin sheath, the activity of fixed enzymes is partly inhibited by the disturbed diffusion of substrate molecules to the enzyme site. Conditions which change the ordered layered arrangement of membranes, such as changes in concentration of ions in the milieu, may either inhibit enzyme activities further or activate the enzymes in accordance with the nature of the change, whether into a water-in-oil or an oil-in-water emulsion. The effects of such changes on enzymic activities in the central nervous system and in myelin will be demonstrated.

UDPG AND UDPGA IN BRAIN AND OTHER ORGANS

Kim Ping Wong and Theodore L. Sourkes

Laboratory of Chemical Neurobiology,
Allan Memorial Institute of Psychiatry, McGill University,
Montreal, Quebec, Canada.

A spectrophotofluorometric method for measuring UDPG and UDPGA is described. The brain is removed immediately after killing the animal, and is then weighed, cut into fine pieces, immersed in boiling water, and homogenized. The supernatant fraction of the centrifuged suspension is adjusted to a tissue:water ratio of 1:1 and 1:4 for the determination of UDPGA and UDPG, respectively. An aliquot of this extract is used directly as a source of UDPGA in a glucoronidating system (0.5 M-Tris buffer, pH 7.8), with harmalol as the co-substrate and the microsomal fraction of guinea pig liver to supply UDP-glucuronyl transferase. The glucuronide so formed is separated from harmalol by TLC (cellulose); the chromatograms are developed with 0.1 N-HCl for 60 min. The spots are readily located under UV illumination because both the aglycone and the conjugate are fluorescent (cf. VILLENEUVE & SOURKES, 1966). The glucuronide is eluted from the adsorbent with 3 ml. of 0.1 N-HCl, and fluorescence is measured in the eluate at 505 nm (activation at 370 nm). Incorporation of the oxidative step in the biosynthesis of DUPGA, catalysed by UDPG dehydrogenase, in the above procedure allows the measurement of UDPG also. This oxidation utilizes a system containing NAD, UDPG dehydrogenase (present in the soluble fraction of guinea pig liver) and 0.1 M-glycine buffer, pH 8.7. UDPG is therefore measured by the difference in extent of glucuronidation before and after dehydrogenation. The concentration of UDPGA in brain of various species is (μmoles/100 gm fresh weight): guinea pig, 2.5; rat, 0.88; mouse, 0.83; hamster, 0.85; and rabbit, 1.00. Corresponding values for UDPG are: 37.1; 29.0; 32.6; 39.2; and 41.4. Concentrations of these uridine nucleotides in other organs, and the effect of hormonal treatments will be reported.

VILLENEUVE, A. & SOURKES, T.L. (1966). Rev. canad. de Biol., 25, 231.

OUTER SEGMENT RENEWAL IN PHOTORECEPTOR CELLS OF THE VERTEBRATE RETINA

Richard W. Young

Department of Anatomy, UCLS School of Medicine, Los Angeles, California, and Département de Biologie, Cen/Saclay.

The photosensitive, visual pigment-containing outer segment of vertebrate photoreceptor cells appears to undergo continual renewal. This conclusion is derived from light microscope autoradiographic studies of the utilization of [^{3}H]methionine in the mature retina in rats, mice and frogs. Shortly after injection, the labelled amino acid becomes concentrated in the photoreceptor inner segment, where the protein synthetic apparatus of the cell is localized. During the following day, the radioactive protein is displaced within the cell, passing through a region of aggregated mitochondria and a ciliated connecting structure to attain the base of the outer segment. Here the labelled material accumulates as a distinct, horizontal band, which is believed to reflect the incorporation of the radioactive protein into newly-forming double-membrane disks. (Work in progress, using electron microscope autoradiography, confirms the incorporation of the labelled protein into the membranous disks at the base of the outer segment). The reaction band is then gradually displaced along the outer segment, towards the apical end of the cell - presumably owing to the continued formation of new disks at the base of that segment. The radioactivity disappears when it reaches the cell junction with the pigment epithelium, which may actively participate in the removal phase of the photoreceptor outer segment renewal mechanism.

In frogs, the outer segment renewal rate is accelerated when the ambient temperature is raised. The renewal rate is slightly elevated in both frogs and rats when the intensity of retinal illumination is markedly increased.

STUDIES ON BRAIN SIALIDASE

V. Zambotti, G. Tettamanti and S. Di Donato

Istituto di Chimica Biologica dell'Università di Milano

An enzyme was isolated from pig brain, which is able to remove sialic acid from polysialogangliosides ('ganglioside-sialidase'). A 1700-fold purified enzyme preparation degraded tetrasialoganglioside $G_{GNT}V$ and trisialoganglioside $G_{GNT}IV$ to disialoganglioside $G_{GNT}III$ and the two disialogangliosides $G_{GNT}III'$ and $G_{GNT}II$ to monosialoganglioside $G_{GNT}I$. The kinetics of enzyme action is now under investigation.

The subcellular localization of the sialidase was studied by homogenization in isoosmotic sucrose and centrifugation in a sucrose density gradient. The activity of the enzyme at different subcellular localizations was observed.

On the basis of (a) the sedimentation pattern, of (b) the action of Na deoxycholate and (c) freezing and thawing treatment, the 'particulate' nature of the enzyme is suggested.

The possible involvement of sialidase in composition and function of brain cell membranes is discussed.

SITE OF INCREASED LABELLING OF RNA IN BRAINS OF MICE DURING SHORT TERM LEARNING EXPERIENCE

John W. Zemp, John Eric Wilson and Edward Glassman

Department of Biochemistry, University of North Carolina School of Medicine, Chapel Hill, North Carolina

Using the double-labelling isotope technique, we have shown increased labelling of RNA of brains of mice undergoing avoidance conditioning (ZEMP et al., 1966). In an attempt to localize the area in which the increased labelling of RNA takes place, brains were dissected and the RNA analysed from each part. Again the double isotope labelling technique was used in which one mouse of a pair received [^{3}H]uridine and the other mouse received [^{14}C]uridine. Thirty minutes after intracranial injection, one of the mice was trained and the other served as a yoke control. The choice of isotope and the choice of animal to be trained were random. Both mice were sacrificed immediately after training and the brains were dissected into olfactory bulbs and tracts (O), hippocampus and temporal lobes (H), frontal lobes (F), parietal lobes (P), cerebellum (C), and diencephalon with anatomically associated areas (D). Corresponding areas from the trained and control mice were combined and homogenized. RNA was isolated from nuclear and ribosomal-polysomal fractions by a modification of the phenol procedure previously described. The radioactivity of the free UMP and of ribosomal and nuclear RNA was determined for each fraction. The average distribution of radioactivity in RNA from the brains of trained or control animals (10 pairs) was less than 5% each in O, P, and H, approximately 10% in F, and 25% in C. Over 50% of the radioactivity in RNA was found in D, where the increase of labelling of RNA in the trained mouse was found. In addition, the percent of the total radioactivity of RNA was greater in D in the trained animal (60% vs 50% in control) with no consistent differences in any other region. We conclude that the increased labelling of RNA during the training experience takes place in the diencephalon or closely associated structures and not in the cortex or cerebellum. A more precise localization of the RNA response will be necessary before it is possible to correlate the response with the aspects of the training experience that might cause the increased labelling.

ZEMP, J.W. et al. (1966). Proc. Nat. Acad. Sci. (USA), 55, 1423.

SEDIMENTATION PROPERTIES OF CEREBRAL RIBOSOMES ACTIVE IN AMINO ACID INCORPORATION

Claire E. Zomzely, Sidney Roberts, Carol Provost and Douglas M. Brown

Department of Biological Chemistry, School of Medicine, and the Brain Research Institute, UCLA, Los Angeles, Calif.

Ribonucleoprotein preparations containing a high proportion of monoribosomes and small aggregates (mixed ribosomes) or preparations consisting primarily of large aggregates (polyribosomes) were prepared from rat cerebral cortex and liver. The cerebral mixed ribosomes incorporated amino acids as actively as hepatic mixed ribosomes. Both preparations exhibited marked increases in amino acid-incorporating activity with poly U. However, sedimentation analyses in the analytical ultracentrifuge and in sucrose density gradients revealed that, unlike hepatic mixed ribosomes, the analogous cerebral ribosomes did not aggregate during incubation in the presence or absence of poly U. Amino acid-incorporating activity of cerebral polyribosomes was comparable to that of hepatic polyribosomes even though the cerebral preparation contained a larger proportion of smaller aggregates which partially dissociated during incubation. By contrast, hepatic polyribosomes did not dissociate under the same conditions. Variations in endogenous RNase activity did not appear to play a role in the differences in ribosomal aggregation observed in cerebral and hepatic ribonucleoprotein preparations. Thus, it appears that cerebral mRNA-polyribosomal complexes are relatively unstable and that smaller ribosomal aggregates may be more active in cerebral protein synthesis than in hepatic systems.

Aided by a research grant from the National Institutes of Health, United States Publs Health Service (GM-12520).

INDEX

ADDENDUM

AMINO ACID METABOLISM IN POSTGANGLIONICALLY DENERVATED SUPERIOR CERVICAL SYMPATHETIC GANGLIA OF THE RAT

Yutaka Nagata and Yasuzo Tsukada

Department of Physiology, Toho University School of Medicine, Tokyo and Department of Physiology, Keio University School of Medicine, Tokyo.

When postganglionic nerves of the cervical ganglion were cut one week prior to isolation, histological examination of the ganglion showed a disappearance of nerve cells by retrograde degeneration and a proliferation of non-neuronal elements. The denervation caused considerable swelling of the ganglion, in which significant increase of both wet and dry weight and rise of water content were involved. The increase of dry weight indicated new formation of solid material due to cellular growth. The amount of DNA increased and ^{32}P incorporation into RNA was raised, although the RNA content per g wet weight remained unchanged in denervated tissue. However, the free amino acid content of the denervated ganglia stayed unchanged in comparison with controls, in spite of the disappearance of nerve cells. This means that the proliferated material consisted not only of connective tissue but also of glial components, since the major amino acids in the denervated tissue were still glutamic and aspartic acids.

The labelling of free amino acids from [U-^{14}C]glucose was mainly in glutamate and aspartate, but not in GABA, both in denervated and in control ganglia. An active participation of neuroglial elements in amino acid metabolism could be inferred.

Preferable uptake of the L-amino acid isomer was shown in isolated control ganglia, but the stereospecificity in amino acid transport was lost after postganglionic denervation. Accordingly, the stereospecificity may possibly belong, not to the glia, but to the neurons. Postganglionically denervated ganglia accumulated glutamate actively almost to the same extent as the controls during periods up to 1.5 hours of incubation. However, the accumulation approached a saturation level within 3 hours, whereas the control ganglia showed a continuous increase of glutamate accumulation during 4.5 hours of incubation. It might be postulated that glutamate from the medium first accumulated in satellite glial cells until their capacity for amino acid accumulation had been fulfilled, and subsequently it was transferred to neighbouring nerve cells showing the stereospecificity.

STEROL METABOLISM IN THE BRAIN OF THE DEVELOPING CHICK EMBRYO

J.F. Weiss*, G. Galli** and E. Grossi Paoletti

Institute of Pharmacology, University of Milan, Italy.

It has been shown previously that desmosterol accumulates in the immature nervous tissue of animals of various species and is also detectable in some experimental and human brain tumours (PAOLETTI *et al.*, 1967). Radioisotopic studies have indicated that desmosterol serves as a major precursor of brain cholesterol in the chick embryo (FISH *et al.*, 1962). In the present study, a further attempt was made to determine the significance of the presence of a desmosterol pool in rapidly developing nervous tissue, as well as to isolate any other sterols which might be present, in order to elucidate the pathway of brain cholesterol biosynthesis.

Labelled mevalonic acid was injected into the yolk sacs of chick embryos after 11 and 18 days of incubation, and the embryos were harvested 4 hours after injection. Sterol acetates, prepared from the unsaponifiable fraction of the brain, were separated on columns containing silver nitrate-silica gel G:celite. The sterol fractions were qualitatively and quantitatively analyzed by gas-liquid chromatography (GLC). Radioisotope incorporation into the sterol fractions was determined and the structures of the major sterol components were elucidated by GLC-mass spectrometry. Small amounts of a number of cholesterol companions, other than desmosterol, were found to be present in the developing brain. There was active incorporation of mevalonic acid into desmosterol, as well as into other sterols that occur in the biosynthetic sequence from lanosterol to cholesterol.

The relationship between brain sterols and those of other tissues of the chick embryo were studied using silver nitrate thin-layer chromatography. Brain and liver were compared to the yolk sac membranes, which can actively incorporate mevalonic acid into sterols other than cholesterol.

*U.S. Public Health Service Postdoctoral Fellowship (No.1-F2-CA-31,598-01) from the National Cancer Institute

**Fellowship from the Commission for Scientific Research of Italian Switzerland.

PAOLETTI, R., FUMAGALLI, R., GROSSI, E. & PAOLETTI, P. (1965). J. Am. Oil Chem. Soc., 42, 400. FISH, W.A., BOYD, J.E. & STOKES, W.M. (1962). J. Biol. Chem., 237, 334.